Environmental Management

Environmental Management

Vijay Kulkarni
and
T.V. Ramachandra
Centre for Ecological Sciences
Indian Institute of Science, Bangalore

Commonwealth of Learning, Canada
Centre for Ecological Sciences, Indian Institute of Science, India
Karnataka Environment Research Foundation, India

First reprint in 2009
Second reprint in 2011
Third reprint in 2014
Fourth reprint in 2015
Fifth reprint in 2018

Environmental Management

Course Writers: Dr. Vijay Kulkarni and Dr. T.V. Ramachandra
Instructional Designer and Editor: Dr. Krishnapillai Murugan

Production Assistant: Mrs. Sherry-Anne Hutson
Content Development Assistants: Saira Varghese K. and Ahalya N.

Content Reviewers:
Dr. Rajasekara Murthy C.
Dr. Giridhar D.P.
Dr. Ramachandra T.V.
Dr. Krishna Alluri

Instructional Design Reviewer: Dr. P.R. Ramanujam
Project Coordinator: Dr. T.V. Ramachandra
Project Manager: Dr. Krishna Alluri

ISBN: 978-81-7993-184-4

Published by
TERI Press
The Energy and Resources Institute
Darbari Seth Block
IHC Complex, Lodhi Road
New Delhi – 110003
India

Tel. 2468 2100 or 4150 4900
Fax 2468 2144 or 2468 2145
India +91 • Delhi (0) 11
E-mail teripress@teri.res.in
Web www.teriin.org

Typeset by TERI Press, New Delhi
Printed in India at Innovative Designers and Printers, New Delhi

Contents

COMMONWEALTH of LEARNING

Commonwealth of Learning

The Commonwealth of Learning (COL) is the only international intergovernmental agency that focusses exclusively on using technology to expand the scope and scale of human learning. It operates on the premise that knowledge is the key to individual freedom and to cultural, social and economic development.

COL helps Governments to develop policies that make innovation sustainable and to build systems or applications that expand learning. It works in partnership with other international and bilateral organisations working on the Millenium Development Goals. These include: the Canadian International Development Agency (CIDA), the UK's Department for International Development (DFID), New Zealand's Agency for International Development (NZAID), the Food and Agriculture Organization of the UN (FAO), the World Health Organization (WHO), the United Nations Educational, Scientific and Cultural Organization (UNESCO), the United Nations High Commission for Refugees (UNHCR), The Joint United Nations Programme on HIV/AIDS (UNAIDS), the New Partnership for African Development (NEPAD), the African Virtual University (AVU), the regional Development Banks and the World Bank.

COL is a small agency. It achieves high impact through its focus on technology; placing special emphasis on open and distance learning (ODL) because of its proven effectiveness.

The Chairman of the Board of Governors is HE, the Hon'ble Burchell Whiteman, OJ, and COL's President and Chief Executive Officer is Sir John Daniel.

www.col.org

Indian Institute of Science

http://www.iisc.ernet.in
http://ces.iisc.ernet.in/energy

The Indian Institute of Science (IISc) was started in 1909 through the pioneering vision of J.N. Tata. Since then, it has grown into a premier institution of research and advanced instruction, with more than 2000 active researchers working in almost all frontier areas of science and technology. IISc is an institute of higher learning and is constantly in pursuit of excellence. It is one of the oldest and finest centres of its kind in India, and has a very high international standing in the academic world as well.

The Institute currently has more than forty academic departments, pursuing research in different areas of science and engineering. The departments belong to one of the two faculties of the Institute, *viz.*, the **Faculty of Science** and the **Faculty of Engineering**. Faculty of Science awards PhD degrees through the Regular Research and Integrated PhD programme. On the other hand, Faculty of Engineering awards ME, MTech and MDes degrees.

The Institute has pioneered advanced education in India, and has been making many significant contributions to frontier areas of research. The number of students in the institute is kept deliberately small in order to focus on quality. It has been able to innovate and introduce (a) new systems of imparting knowledge; and (b) educational reforms such as offering courses under unit system.

Preface

The life supporting systems of the planet's biosphere are being threatened due to deforestation, destruction of habitats, over use of energy resources and environmental pollution. Changes in the earth's climate, decline and deterioration of natural resources, the accumulation of waste products, soil exhaustion and the destruction of ecosystems, are already apparent. Anthropogenic activities coupled with the burgeoning human population are responsible for the loss of a large number of life forms; numerous important plant and animal species are on the verge of extinction, while others are threatened or vulnerable. In order to bring about sustainable resource conservation and management, it is essential to adopt several different approaches for managing the ecosystem and biodiversity. To arrest the process of degradation and species loss, it requires specialised solutions and an understanding of ecological processes. Human beings have been interested in ecology since the beginning of civilization. Even our ancient scriptures have included practices and values related with ecological and environmental conservation. India is one of the mega diversity countries and a Vavilovian centre of origin and diversification of cultivated plants, and endowed with rich traditional knowledge system. Currently, it is even more critical than ever before for the humankind as a whole to have a clear understanding of environmental concerns and to follow sustainable development practices. The need for sustainable development is a key to the future of humankind. The degradation of our environment is linked to continuing problems of pollution, loss of forest, solid waste disposal, and issues related to economic productivity and national as well as ecological security. The increasing levels of global warming, the depletion of the ozone layer and a serious loss of biodiversity have also made everyone aware of growing environmental concerns. However, it is imperative to maintain a balance between the capacity of the environment and the quantum of sustainable utilisation. This is only possible by understanding the environment in its totality and the principles of its scientific management. **Environmental management** has gained momentum in recent years with the initiatives focussing on managing environmental hazards and preventing possible disasters.

The United Nations Conference on Environment and Development held in Rio De Janero in 1992, and the World Summit on Sustainable Development at Johannesburg in 2002 have drawn the attention of people around the globe to the developing condition of our environment. In order to achieve the goals of sustainable development, people need to become aware of the environmental issues and acquire background information to enable them to make and influence decisions. Environmental education is thus concerned with attitude towards, and decisions about environment quality; with informed management of resources; and with the ethical considerations that relates to these. Recognising the importance of environmental education at all levels, the Hon'ble Supreme Court ruled that a course on Environment be made mandatory at all levels of education to sensitise the youth to environmental issues and concerns.

The declaration of the decade for Education for Sustainable Development (ESD) beginning in 2005, by the United Nations has provided further impetus. The goal is to create a sustainable world through active participation of citizens. Thus, ESD is seen as a process that develops vision, builds capacity, and empowers to make changes in human societies. Education has a pivotal role to play in achieving a sustainable economy and society. The cognitive and cultural separation of 'ecology and environment' from the human enterprise, has led to large scale degradation and depletion of natural resources. The guiding ideology needed to learn and teach sustainability is an ideological orientation that emphasises conserving cultural values, beliefs, and practices that contribute to sustainable relationships with the environment. Perhaps the best way to see if institutions of education have begun to develop ecologically is to determine whether or not they acknowledge, in their structure, pedagogy and curriculum.

Environmental management would help to recognise the importance of investigating and managing the environment within the context of human influences, incorporating an examination of economics, culture, political structure, and social equity as well as natural processes and systems. Environmental Management Course cultivates the ability to recognise uncertainty, envision alternative scenarios, and adapt to changing conditions and information. These knowledge, skills, and mindset translate into a citizenry that is better able to address its common problems and take advantage of opportunities, whether environmental concerns are involved or not. Environmental issues make better sense, when one can understand them in the context of one's own cognitive sphere. Environmental education focussing on real-world contexts and issues often begins close to home, encouraging learners to forge connections with and understand their immediate surroundings. The awareness, knowledge, and skills needed for these local connections and understandings provide a base for moving out into larger systems, broader issues, and a more sophisticated comprehension of causes, connections, and consequences. Essentially, EM represents the management of various activities, including environmental planning, conservation of resources, environmental status evaluation and environmental legislation and administration, and focusses more on implementation, monitoring, auditing, practice and real-world issues than on theoretical planning.

Environmental management (EM) thus has become one of the most used terms in recent times. But, what exactly does the term mean and entail? Different individuals belonging to different disciplines approach EM differently and therefore it is difficult to find a single, comprehensive view of EM. However, we will discuss in detail the various issues relating to EM in the ten Units that constitute this Course. We will first discuss the fundamentals of EM and ecosystem in Unit 1 and, subsequently, the various environmental policies, legislations and international treaties in Unit 2. In Units 3 and 4, we will deal with the concept of environmental impact assessment (EIA) and the preparation of EIA report, respectively. We will discuss the methodology and processes of environmental auditing in Unit 5. In Unit 6, we will introduce you to life cycle assessment (LCA) in the context of EM. In Unit 7, we will explain the various EM system standards. We will then take up the issues and techniques relating to EM in Unit 8. Finally, we will discuss environmental design and economics in Units 9 and 10, respectively.

T.V. Ramachandra

UNIT 1

Principles of Environmental Management

STRUCTURE

OVERVIEW

Due to unplanned developmental activities as well as ever-increasing population, which has caused enormous strain on the environmental resources, societies across the world face several problems of environmental degradation. However, it is imperative to maintain a balance between the capacity of the environment and the quantum of sustainable utilisation. This is only possible by understanding the environment in its totality and the principles of its scientific management.

Environmental management (EM) has become one of the most used terms in recent times. But, what exactly does the term mean and entail? Different individuals belonging to different disciplines approach EM differently and therefore it is difficult to find a single, comprehensive view of EM. However, we will discuss in detail the various issues relating to EM in the ten Units that constitute this Course.

We will first discuss the fundamentals of EM and ecosystem in Unit 1 and, subsequently, the various environmental policies, legislations and international treaties in Unit 2. In Units 3 and 4, we will deal with the concept of environmental impact assessment (EIA) and the preparation of EIA report, respectively. We will discuss the methodology and processes of environmental auditing in Unit 5. In Unit 6, we will introduce you to life cycle assessment (LCA) in the context of EM. In Unit 7, we will explain the various EM system standards. We will then take up the issues and techniques relating to EM in Unit 8. Finally, we discuss environmental design and economics in Units 9 and 10, respectively.

Against this background, let us now discuss Unit 1. We will begin this Unit by introducing you to the scope and definition of EM. We will then discuss the fundamentals of EM and ecosystem. We will also touch upon some of the tools in use to address environmental problems; discuss the role of the participants in managing the environment and explain the concept of environmental ethics. We will close the Unit by outlining the current environmental concerns in India.

LEARNING OBJECTIVES

After completing this Unit, you should be able to:

- explain the fundamentals of EM and ecosystem;
- identify the tools in use for EM;
- explain the role of professionals and ethics in EM.

1.1 INTRODUCING ENVIRONMENTAL MANAGEMENT (EM)

The natural environment, harbouring myriad life forms and ecosystems, is an outcome of millions of years of evolution and mutation.

We depend on the environment to meet our basic requirements such as food, fuel, fibre, fodder, minerals and vital support systems (e.g., water and air). However, due to our continuous indulgence in the exploitation of nature for immediate gains, we contribute to environmental degradation and depletion of natural resources, and this should be arrested. It is, therefore, essential that we understand the function and interaction of physical and biological elements of the environment and apply this knowledge for sound management programmes to conserve the natural resources we are endowed with. Sustainable management of renewable resources in industry, agriculture, recreation, health, forestry, fisheries, education or urban development is, therefore, critical. The conservation of resources and maintenance of the quality of ecosystem requires as much imaginative and innovative technology as the active participation of the public.

Essentially, EM represents the management of various activities, including environmental planning, conservation of resources, environmental status evaluation and environmental legislation and administration, and focuses more on implementation, monitoring, auditing, practice and real-world issues than on theoretical planning.

1.1.1 Definition and Scope

Different people define environmental management differently, depending on the context in which they work. For the purposes of this Course, let us consider the following two views:

(i) The process of allocating natural and artificial resources so as to make optimum use of the environment in satisfying basic human needs at the minimum, and more if possible, on a sustainable basis (Jolly, 1978).

(ii) A generic description of a process undertaken by systems-oriented professionals with a natural science, social science, or less commonly, an engineering, law, or design background, tackling problems of the human altered environment on an interdisciplinary basis from a quantitative and/or futuristic viewpoint (Dorney, 1989).

Some of the characteristics of EM include the following:

- It is often used as a generic term.
- It supports sustainable development.
- It deals with a world affected by human beings.
- It demands a multidisciplinary or interdisciplinary approach.
- It has to integrate different development viewpoints.
- It seeks to integrate science, social science, policy-making and planning.
- It recognises the desirability of meeting and, if possible, exceeding basic human needs.
- The time-scale involved extends beyond the short-term and concerns range from local to global.
- It should show opportunities as well as address threats and problems.
- It stresses stewardship, rather than exploitation.

In simple terms, environmental managers need to interface with ecology, economics, law, politics, people, etc., while planning developmental activities.

The fact that much of these activities at a given point in time and space are likely to have wide and long-term impacts makes it desirable for development to be managed and coordinated at all levels: regional, national and international. This implies that an environmental manager must 'think globally and act locally' and adopt a long-term holistic outlook.

1.1.2 Goals of EM

The goals of EM, which is an approach for environmental stewardship integrating ecology, policy-making, planning and social development, include:

- preventing and resolving environmental problems;
- establishing limits;
- establishing and nurturing institutions that effectively support environmental research, monitoring and management;
- identifying threats and opportunities;
- sustaining and, if possible, improving existing resources;
- improving the quality of life;
- identifying environmentally sound technologies or policies.

Thus, EM demands *scoping*, i.e., deciding goals and setting limits on efforts.

Having touched upon the definition and characteristics of EM, we will discuss some of the crucial issues relating to EM such as the population growth, sustainable development and EM tools. But, let us first work out Learning Activity 1.1.

LEARNING ACTIVITY 1.1

List some of the goals of environmental management.

Note:

a) Write your answer in the space given below.

b) Check your answer with the one given at the end of this book.

____

____

____

____

____

____

____

____

____

1.1.3 Implications of Population Growth

The world population is estimated to be over six billion. The human population is continuously increasing. In the last 10,000 years, the population of the world has increased over a 1000-fold and much of that change has occurred in the last century. This phenomenal growth in population has put pressure on the means of subsistence, throwing it out of balance with the environment.

The interaction between population and environment is very complex and dynamic. Admittedly, however, we are yet to understand this complex interrelationship at its micro-level and its spatio-temporal ramifications over a region or the globe in its entirety. What is, nevertheless, certain is that as the population grows, the level of consumption of natural resources and production of wastes proportionately increases (Harrison, 1990). And, we know that the environment is constantly changing, due to human activities, leading to such problems as soil erosion, floods, droughts, climate changes, desertification and general degradation of the environment. In a finite world, there are *limits*, and indeed complex environment-population linkages and feedback.

Sustainable development is, therefore, necessary to sustain the quality of life without exceeding the environmental limits. We will study this, next.

1.1.4 Need for Sustainable Development

The great surge in development and technology over the last two centuries has contributed to the increase in quantities of chemicals sufficient to damage natural systems on a grand scale, more often than not, involving the whole world. Acid rain, desertification, destruction of species, greenhouse modification and ozone layer depletion are well known examples of these impacts.

Protection of the natural ecosystem has long-term benefits for humans in utilitarian terms through maintenance of gene pools, biodiversity and other potentially useful factors and, in spiritual terms, through living in harmony with nature. The ecosystem's intrinsic values and rights, regardless of human needs, therefore, should be taken into account apart from considering it a resource to be exploited for human settlement, food and energy production.

Sustainable development, therefore, is imperative. It is defined as a pattern of social and structured economic transformation (i.e., development), which optimises the economic and societal benefits available in the present, without jeopardising the likely potential for similar benefits in the future. A primary goal of sustainable development is to achieve a reasonable and equitably distributed level of economic well being that can be perpetuated continuously for many human generations. This implies using natural resources in a renewable manner that does not eliminate or degrade them, or diminish their usefulness for future generations. It further implies using non-renewable (i.e., exhaustible) mineral resources in a manner that does not unnecessarily preclude easy access to them by future generations.

Three of the several definitions that capture these aspects are given below:

(i) *Development that meets the needs of the present without compromising the ability of future generations to meet their own needs* (WCED, 1987).

(ii) *Sustainable development ensures that the maximum rate of resource consumption and waste discharge for a selected development portfolio would be sustained indefinitely, in a defined planning region, without progressively impairing its bio-productivity and ecological integrity. Environmental conservation, therefore, contrary to general belief, accelerates rather than hinders economic development. Therefore, the development plans have to ensure:*

- *sustainable and equitable use of resources for meeting the needs of the present and future generations without causing damage to environment;*
- *to prevent further damage to our life-support systems;*
- *to conserve and nurture the biological diversity, gene pool and other resources for long-term food security* (MoEF, 1999).

(iii) *The primary objective of sustainable development is to reduce the absolute poverty of the world's poor through providing lasting and secure livelihoods that minimise resource depletion, environmental degradation, cultural disruption and social instability* (Barbier, 1987; *http://sdnp.delhi.nic.in*).

Thus, the doctrine of sustainability provides a mechanism for developments to occur in harmony with environmental protection and enhancement. It presents a challenge to technologists to manage construction and manufacturing procedures so as to cause no net environmental harm and to preserve environmental resources for future generations.

Some of the approaches to EM to achieve sustainability include the following:

- **Ad hoc approach:** This means an approach developed in reaction to a specific situation.
- **Problem-solving approach:** This refers to an approach that follows a series of logical steps to identify problems, needs and solutions.
- **Systems approach:** This approach focuses on ecosystem (e.g., mountain; high latitude; savannah; desert; island; lake, etc.), agro-ecosystem, etc.
- **Regional approach:** This is based on ecological zones or biogeophysical units such as watershed, river basin, coastal zone, islands, etc.
- **Specialist discipline approach:** This refers to the approach often adopted by professionals like air quality management, water quality management, land

management, environmental health, urban management, conservation area management, etc.

- **Voluntary sector approach:** This represents the approach NGOs (non-governmental organisations) encourage and/or support.

It is necessary to have a basic understanding of ecology to adopt the approaches mentioned. We will explain the relationship between environment and ecology in Section 1.2.

LEARNING ACTIVITY 1.2

Write three most important driving forces to be considered in planning for sustainable development.

Note:

a) Write your answer in the space given below.
b) Check your answer with the one given at the end of this book.

1.1.5 EM Tools

One of the outcomes of the three decades of policy development since 1970 has been the evolution of the techniques for the analysis and management of environmental effects. The tools considered in this Course include:

(i) Environment assessment (EA).
(ii) Economic assessment, usually through cost-benefit analysis (CBA).
(iii) Environment impact statements (EIS).

(iv) Environment audits.
(v) Waste minimisation programmes and environmental management systems.
(vi) Life cycle assessments (LCA).
(vii) Environmental design (ED).

These tools are indispensable as far as environmental management is concerned. But, rather than being comprehensive, self-contained procedures, these tools are really aids that can be adapted for various situations. Legislative and bureaucratic frameworks for environmental management differ considerably, not only between nations, but also among states or regions within countries. Some procedures fit within others, for example, an environment impact statement is a tool or instrument, used within a more general environment assessment process. The methods are contributions to decision-making, rather than complete decision-making procedures. You will learn these tools in greater detail from Unit 3 onwards.

1.2 ECOSYSTEM CONCEPTS

Ecology consists of both biotic or living and abiotic or non-living factors. The sum of all these living and non-living factors makes the environment of an organism. The place where an organism lives is called its habitat/milieu. In fact, habitat is a part of the earth's surface where life originates and develops according to geographical conditions.

Figure 1.1 illustrates the interrelationship among ecological elements.

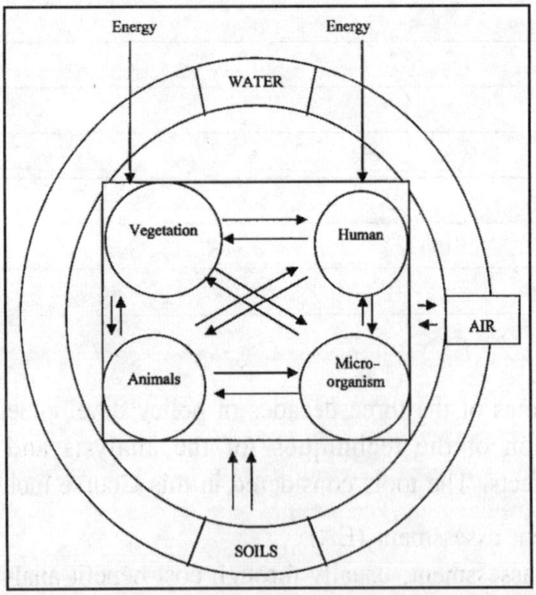

Figure 1.1 Interrelationships of Ecological Elements

The biotic and abiotic elements, as shown in Fig. 1.1, are interrelated and function in an orderly manner as a definite system. The cycling of materials between living and non-living factors is an ecological system or ecosystem. An ecosystem, thus, is the basic functional unit and includes both organisms and abiotic environment, each influencing the properties of the other and both necessary for maintenance of life on earth.

1.2.1 Structure of an Ecosystem

From the structural point of view, all ecosystems consist of the following:

(i) **Biotic components:** On the basis of their nutritional (trophic) status, organisms in an ecosystem are broadly divided into **autotrophs** and **heterotrophs**. Autotrophs are organisms that can produce their own food such as green plants (with chlorophyll) and certain bacteria, which obtain their energy from the sun and through photosynthesis, produce food. These are usually known as *producers*, since these organisms produce food for all the other organisms. Heterotrophs, known as *consumers*, depend directly or indirectly upon the autotrophs for their food. They are grouped into:

- **Primary consumers:** These include herbivorous animals (e.g., rodents, cows, deer, goats, buffaloes, etc.), which depend on green plants for their food.
- **Secondary consumers:** These include carnivores and omnivores (e.g., sparrows, crows, foxes, wolves, dogs, etc.)
- **Tertiary consumers:** These are top carnivores (e.g., lions, tigers, hawks, vultures, etc.), which prey upon herbivores and omnivores.
- **Decomposers and transformers:** These are the living components of the ecosystem such as fungi and bacteria. They decompose the dead remains of the producers (i.e., autotrophs) and consumers (i.e., heterotrophs) and convert the complex organic substances into simple organic compounds. These are further broken down by bacteria and converted into inorganic forms, which are taken up by green plants (Mukherjee, 2002).

(ii) **Abiotic components:** Abiotic components of the ecosystem include a variety of organic compounds and basic inorganic elements and compounds such as soil, water, oxygen, calcium, carbonates and phosphates. The physical factors such as moisture, wind direction and solar radiation also form part of abiotic components. The amount of non-living components (e.g., calcium, phosphorous and nitrogen) present at any given time is known as the *standing state*.

Nutrient Cycling

Nutrients are atoms and molecules that are essential for the growth of plants and trees. Examples of nutrients that are important to plants are carbon (C), nitrogen (N), magnesium (Mg), calcium (Ca), iron (Fe) and phosphorus (P). Note that nutrients are not the same as energy, which comes originally from the sun and is converted, via photosynthesis, into sugars that feed the plant. The fundamental difference between energy and nutrient transfer is that the pattern of nutrient transfer is basically circular or cyclical.

Different nutrients are important to different chemical processes within and between plant cells. For example, plants use nitrogen to synthesise proteins, DNA and many other molecules. Magnesium is a component of chlorophyll, i.e., the molecule that helps plants capture solar energy.

Cycling ensures that nutrients do not get stuck somewhere in the soil. For instance, when a plant dies, all its nutrients end up in the soil. Bacteria and fungi break down the plant tissue and free up the nutrients, which other plants growing in the same soil consume. Recyclers (mainly fungi and bacteria) are important in maintaining availability of nutrients in all ecosystems. Physical processes such as air and water movement influence the fate of these nutrients to variable extents. The cycles also differ in scale from local to global, and each is increasingly subject to human impact—deliberate or accidental. The major cycles occurring in nature include the following:

- **Water cycle:** This essentially involves physical processes of precipitation, evaporation and water flow with solar energy as the driving force. However, organisms and especially plants modify evaporation and precipitation. The presence of water in oceans and the atmosphere makes this a global cycle.
- **Carbon cycle:** This is also global but organic processes (e.g., photosynthesis and respiration) play a more dominant role than in the water cycle.
- **Nitrogen cycle:** This is perhaps the most complex in that so many organisms are involved in the transformations. Essentially, the only way nitrogen finds its way into the food chain is through nitrogen-fixing bacteria. Note that nitrogen is the limiting nutrient in many ecosystems.
- **Phosphorus cycle:** This is simpler than the nitrogen cycle and does not depend on a single class of organisms. Phosphorus is widely present in the earth's crust, although it may often be insoluble and unavailable to plants.

Usually, decomposers play a vital role in nutrient cycling in returning small molecules and ions back from the higher trophic levels to the soil, water or air, where they are available for reuse by the primary producers. However, there are certain factors that disrupt nutrient cycling including the following:

- Acid rain can change the chemistry of soil and change the nature of certain nutrients.
- Deforestation can strip away all the trees and lead to erosion, which washes away all the rich soil that is full of nutrients for future plants.
- Natural disturbances such as fire or drought can interrupt the normal cycling of nutrients caused by movements of water.

1.2.2 Biomass, Energy and Energy Flow

Many factors can limit the growth of trees and other forms of life, but the ultimate limit on production of organic matter is energy flow. Before determining the maximum possible production of organic matter of any kind, it is necessary to understand the basic concepts of biological or biomass production, energy and energy flow. We will study these, next.

Biological Production

Biological production involves the capture of usable energy and the production of organic compounds in which the energy is stored. This includes all living beings and their products. Once an organism has obtained a new organic matter, it can use the energy to move and to produce and reproduce new compounds. The use of energy in organic matter in both heterotrophy and autotrophy is accomplished through respiration. In respiration, an organic compound is combined with oxygen to release energy and produce carbon dioxide and water.

There are three measures of production, viz., biomass, energy content and carbon content. Biomass is usually measured as the amount of total organic matter per unit surface area of the earth, for example g/m^2 or metric tons per hectare, MT/ha. The change in biomass over a given period of time is called net production.

There are three steps in the production of biomass and its use, as a source of energy, by autotrophs. First, an organism produces organic matter within its body; next, it uses some of this new organic matter as a fuel in respiration and finally, some of the newly produced organic matter is stored for future use. That is to say:

Net production = Gross production – Respiration

For example, the net production of the tree is energy contained in what is left at the end of the year and includes new wood laid down in the trunk, new buds that will develop into leaves and flowers the next year, and new roots. Net production results in the increase of total weight, energy content and stored carbon (Botkin and Keller, 1995).

Energy in Ecological Systems

Energy is defined as the ability to do work. The law of entropy or thermodynamics describes the behaviour of energy. The first law of thermodynamics, or energy conservation, states that energy may be transformed from one type into another but is neither created nor destroyed. According to the second law of thermodynamics, no process involving an energy transformation will spontaneously occur, unless there is a degradation of the energy from a concentrated form into a dispersed form.

Organisms, ecosystems and the entire biosphere possess the essential thermodynamic characteristics. In other words, these can create and maintain a high state of internal order, or a condition of low entropy (i.e., a low amount of disorder or unavailable energy in a system). *Low entropy* is achieved by continually and effectively dissipating energy of high utility (e.g., light or food) to energy of low utility (e.g., heat). Ecosystems and organisms are, accordingly open, non-equilibrium thermodynamic systems that exchange energy and matter with the environment continuously to decrease internal but increase external entropy.

Energy Flow

Energy flow is the movement of energy through an ecosystem – from the external environment through a series of organisms and back to the external environment. It is one of the fundamental processes common to all ecosystems. Figure 1.2 illustrates the energy flow within an ecosystem.

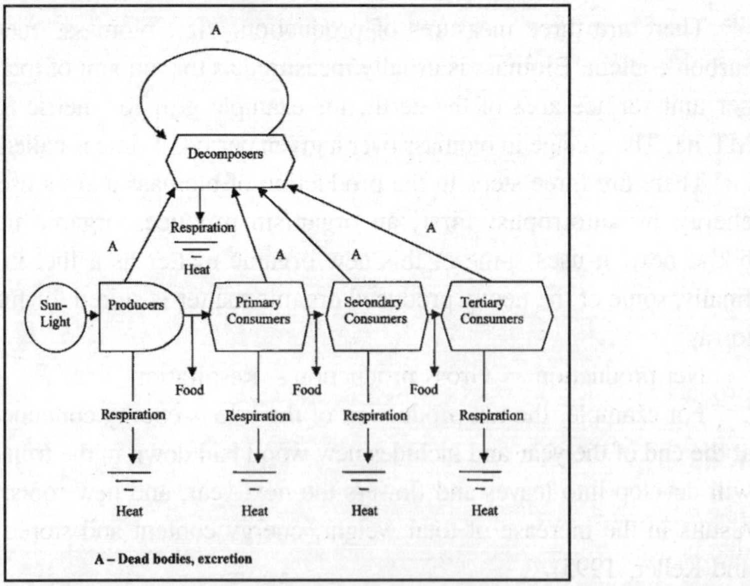

Figure 1.2 Energy Flow Within an Ecosystem

Having studied the concepts of biomass production, energy and energy flow, let us now understand the general relation between biomass (B) and net production (NP). The equations used for measuring biomass are:

- $B_2 = B_1 + NP$ (1)
- $NP = B_2 - B_1$ (2)

where B_2 is the biomass at the end of the time period; B_1 is the amount of biomass at the beginning of the time period and NP is the change in biomass during the time period. The general production equations are:

- $GP = NP + R$ (3)
- $NP = GP - R$ (4)

where *GP* is the gross production; *NP* the net production and *R* is respiration.

Note that the average of the energy in vegetation is approximately 21 kJ/g (1 kJ = 1000 J = 0.24 kilocalorie). The kilocalorie (kcal) is the amount of energy required to heat a kilogram of water by one degree Celsius (from 15° C to 16°C).

To understand these equations better, let us now work out a problem.

Consider a forest with an annual increase in forest biomass as 1060 g/m^2 per year. This figure is the annual net production rate or net change in biomass. If the biomass at the beginning of the year was measured at 9700 g/m^2, so that the new biomass in the forest at the end of one year would be:

$$B_2 = 9700 \text{ g/m}^2 + 1060 \text{ g/m}^2 = 10{,}760 \text{ g/m}^2$$

Since, on an average, 1 g of completely dried vegetation contains about 21 kJ of energy, the net production of this forest at the end of one year in terms of energy would be:

$$NP = 21 \text{ kJ/g} \times 1060 \text{ g/m}^2\text{/year} = 22,260 \text{ kJ/m}^2\text{/year}$$

About 45% of vegetation in dry matter is carbon, so we can estimate the change in carbon during the year as:

$$NP = 0.45 \times 1060 \text{ g/m}^2 = 477 \text{ g/m}^2$$

The total amount of carbon after one year would be:

$$B_2 = 0.45 \times 9700 \text{ g/m}^2 + 0.45 \times 1060 \text{ g/m}^2 = 4842 \text{ g/m}^2$$

This example demonstrates how we can calculate the energy and carbon production from the biomass production.

1.2.3 Food Chains, Food Webs and Trophic Levels

The transfer of energy from the autotrophs (i.e., plants) through a series of organisms that consume and are consumed is called the **food chain** as illustrated next:

Marsh grass → Grasshopper → Bird → Hawk

At each transfer, a proportion (often as much as 80% or 90%) of the potential energy is lost as heat. Therefore, the shorter the food chain, the greater the energy available to that population. Food chains are of two basic types, and these are:

(i) **grazing food chain**, which, starting from a green plant base, goes to grazing herbivores and on to carnivores;
(ii) **detritus food chain**, which goes from non-living organic matter into micro organisms and then to detritus-feeding organisms (detritivores) and their predators.

Food chains are not isolated sequences; they are interconnected. The interlocking pattern is often spoken of as the **food web**.

In complex natural communities, organisms whose nourishment is obtained from the sun, by the same number of steps, are said to belong to the same trophic level. Thus, green plants (the producer level) occupy the first trophic level, herbivores the second trophic level (the primary consumer level), primary carnivores the third level and secondary carnivores the fourth level (the tertiary consumer level). Note that this trophic classification is based on function and not of species as such. In other words, a given species population may occupy one or more than one trophic level according to the source of energy actually assimilated. The energy flow through a trophic level equals the total assimilation (A) at that level, which, in turn, equals the production (P) of biomass and organic matter plus respiration (R).

1.2.4 Ecosystem Restoration and Recovery

Recent studies show that ecosystems, even when undisturbed by people, are not static but change over time. Studying the patterns of the change in ecosystems in response to natural or human disturbance is an important tool for learning how to restore lands damaged by people.

Natural areas are subject to disturbances of many kinds. These disturbances are not always or even usually human-induced. For example, natural disturbances such as storms and fires have always been a part of the environment. Such disturbances have existed for so long that animals and plants have adapted to them to such an extent that the landscape actually benefits in the long run. When fires occur at natural rates and intensities, there are some beneficial effects. For example, trees in unburnt forests appear more susceptible to insect outbreaks and disease. Thus, recent ecological research suggests that wilderness depends on change and that ecological succession and disturbances are continual processes that keep the landscape dynamic.

There are two kinds of succession – primary and secondary. While primary succession is the initial establishment and development of an ecosystem, the secondary succession is a reestablishment of an ecosystem. Forests that develop on new lava flows or at the edge of a retreating glacier are examples of primary succession. In secondary succession, there are remnants of a previous biological community, including such things as organic matter and seeds in the soil of a forest. By contrast, in primary succession such remnants are non-existent or negligible. A forest that develops on an abandoned pasture or one that grows after a hurricane, flood or fire is an example of secondary succession.

One feature of succession is that the biological community changes. In early stages, biomass increases and so does the diversity of life forms. In middle stages of succession, there are many species and sizes of trees and a variety of other patterns that occur. Gross and net productions change during succession in that while the gross production increases, the net production decreases. Chemical cycling also changes and the organic material in the soil increases, as does the amount of chemical elements in the soil and trees.

What the foregoing discussion shows is that natural processes could restore a damaged habitat into its previous or even into an improved state and that the environment overall is self-sufficient. However, the balance that is maintained is delicate. Removal or addition of certain factors, for example, can disrupt this balance and the result is an imbalance in the ecosystem. We will touch upon this, next.

1.2.5 Effect of Imbalance on the Ecosystem

In most cases, a habitat is in balance, with the right number of producers, herbivores and carnivores to keep it going. However, changes such as the introduction or removal of predators in a habitat, can throw the environment out of balance. Disturbances to biotic factors and abiotic factors will lead to environmental resistance. This will lead to an exponential decrease in population of ecosystems that will cause a high extinction rate of biodiversity. If this condition is severe, the ecology of the ecosystems will be permanently damaged.

The imbalance in the ratio of producers, consumers and decomposers also distorts the pattern of energy flow through the food chain and food web, disrupting the *ecological food pyramid*. When the ecological food pyramid is disrupted, insufficient consumption of food leaves organisms deprived of energy, thus affecting the organism's metabolism. This leads to mass starvation and mortality in world population. Also, this leads to the extinction of particular species, while their predators dominate resulting in ecological niche in ecosystems change.

Having studied some of the basic concepts of and the importance of preserving the ecosystem and the environment, let us now examine the tools necessary to attain a sustainable approach to environmental management.

LEARNING ACTIVITY 1.3

List the various environmental management tools mentioned in this Unit.

Note:

a) Write your answer in the space given below.

b) Check your answer with the one given at the end of this book.

1.3 PARTICIPANTS IN EM

Adam (1990) identified the following two groups involved in environment and development:

(i) People or governments who are uninformed of the implications of development, or who are unable to voice their views adequately and affect change.

(ii) Consultants, scientists, economists and bankers and those bent on riches or blinkered by concern for sovereignty, religion or national security.

In any given environmental management situation there are likely to be a number of different perspectives, and hence the various possible responses. An environmental manager has to grasp the sum total of the perspectives and try to avoid conflicts between participants and minimise damage to the environment. The participants of environmental management can be categorised as existing users, groups seeking change, groups with little control, the public, facilitators and controllers. In what follows in this Section, we will discuss these categories.

1.3.1 Existing Users

This category refers to those who currently use land or other resources and, those using the environment or resources usually evolve rights and develop management skills. However, problems arise where unwritten traditional strategies and rights break down or get usurped, typically, by incoming migrants and settlers, urban elites or powerful commercial organisations. Worldwide, the expropriation of common resources from traditional users has become a problem (The Ecologist, 1993).

There has, however, been a growing practice of seeking to consult and involve local people (i.e., indigenous groups) in environmental management, and to understand and make wider use of indigenous knowledge (Klee, 1980). EM can learn a lot from the study of local people's livelihood strategies.

Since the 1975–85 UN Decade for women, there has been an interest in studying the role of women in EM. According to the different perspectives adopted, these studies take different shapes as under:

- **Women, environment and development:** This focuses on women as having a special relationship with the environment as its users and managers.
- **Gender and development:** This seeks gender as a key dimension of social difference affecting people's experiences, concerns and capabilities.
- **Women in development:** This focuses on reasons for women's exclusion or marginalisation from decision-making and receipt of the benefits of development.

1.3.2 Groups Seeking Change

This category consists of governments (with conflicting demands from various ministries or policy-makers), commerce (e.g., national, local, multinational companies, etc.), individuals seeking personal gain or seeking to change the prevailing situation, international agencies, NGOs, media, academies, etc. It is probably the exception to the rule for special-interest groups not to control policymaking and development, although a few do so with the aim of improving environmental care. The environmental manager should be vigilant for such control, and seek to reduce it if it acts against environmental quality. When environmental management involves more than one country, which is often the case, negotiation skills are at least as important as access to technology, knowledge and management strategies (Vogler and Imber, 1995).

1.3.3 Groups with Little Control

This category consists of the poor with no option but to over-exploit what is available without investing in improvement; refugees, migrants, relocates, eco-refugees (i.e., those forced to move or marginalised so that they change the environment to survive), workers in industry/mining, etc., who face health and safety challenges, while carrying out changes. Many identify poverty alleviation and environmental care as two challenges for those in charge of development. These two issues are closely related, although linkages are often unclear and complex.

The poor, it is often claimed, degrade their environment in an effort to survive– a trap of poverty. Getting people out of poverty may be important for protecting the environment, but environmental managers must consider each local case to be sure of causes. For example, the causes of environmental degradation in urban areas may lie with policies affecting agriculturalists hundreds of kilometres away, causing them to migrate and increase city population. There are also situations where there is likely to be poverty-related environment stress: cities where population growth is outstripping employment and infrastructure; marginal, often vulnerable land where people have relocated, areas where traditional livelihood strategies are degenerating.

1.3.4 The Public

People, who are affected as bystanders, may wish to develop, conserve or change practices (if aware of what is happening) and those out of global concern form part of this category. The public usually consists of more than one group of people who probably have different, perhaps conflicting, views and goals with powerful groups dominating the situation. Environmental managers must, therefore, establish the needs of the weak and ensure that they are not ignored, yet work with the influential.

Sustainable development strategies need to be designed to fit local conditions and to be co-ordinated to ensure that one locality does not conflict with another. Environmental management should act as mediator and catalyst to develop collaborative approaches (Selin and Chavez, 1995). Advantages of public involvement in environmental management are as follows:

(i) The public may be able to provide advice on management considering local conditions.
(ii) Often planning and management should be more accountable and more careful.
(iii) Fears and opposition to management may be reduced, if people are informed.
(iv) The communication gap between the experts and locals can be reduced.

1.3.5 Facilitators

This category consists of funding agencies, consultants, planners, workers including migrants affected by health and safety issues, etc.

Funding bodies can support environmentally desirable developments or withhold money until proposals are modified to meet required standards. Starting with the World Bank in the early 1970s, most funding bodies have developed environmental management units, guidelines and manuals (Turnham, 1991). There is also a huge diversity of bodies conducting research aimed at improving environmental management: universities, private research companies, independent international research institutes and UN or UN-related agencies. Most research is applied in response to perceived needs, but some is anticipatory and warns about possible threats and potentially useful strategies.

1.3.6 Controllers

This category consists of government and international agencies, traditional rulers and religions, planners, law, consumer protection bodies and NGOs (including various green/environmental bodies), trade organisations, media, concerned individuals, academics, global opinion, environmental managers, etc.

LEARNING ACTIVITY 1.4

List three most important activities of your profession that affect environment.

Note:

a) Write your answer in the space given below.
b) Check your answer with the one given at the end of this book.

____

____

____

____

____

____

____

____

____

____

____

____

The skills of environmental managers and ecologists are vital to determine the best strategies for the survival of fauna and flora and to organise sustainable land and resource use. NGOs have become important watchdogs of corporate, government and special-interest group activities. They have a multifaceted role: lobbying at international meetings and at national government level; media campaigning to increase public awareness and empowerment; fund-raising for environmental management, conservation and environmental education; researching environmentally sound strategies and approaches; acting as ginger groups to identify environmental problems and fight for their control, etc.

A proper environment is essential to sustain all renewable resources: food, fibre, wood, air, water and even our health. However, the debate over the ethical values attached to the environment is as old as civilisation. We will discuss this in Section 1.4. But, before we do so, let us work out Learning Activity 1.5.

LEARNING ACTIVITY 1.5

What advantages do you see for yourself as a professional in becoming acquainted with the field of environment management?

Note:

a) Write your answer in the space given below.

b) Check your answer with the one given at the end of this book.

1.4 ETHICS AND THE ENVIRONMENT

In the 1970s, philosophers began to formulate a new field called environmental ethics, a study concerned with the value of the physical and biological environment. The focus of this study contrasts with traditional ethical studies, which had to do with the relationships among people. There are utilitarian, ecological, aesthetic and moral reasons for placing a value on the environment. The major factors that necessitate environmental ethics are:

(i) **New effects on nature:** Our modern technological civilisation affects nature greatly, hence we must examine the ethical consequences of these new actions.

(ii) **New knowledge about nature:** Modern science demonstrates how we have changed and are changing our environment in ways not previously understood, thus raising new ethical issues. For example, until the past decade, few people believed that human activities could be changing the earth's global environment. Now, however, scientists believe that burning fossil fuels and clearing forests have changed the amount of carbon dioxide in the atmosphere, and this causes changes in our climates, and hence the need for global action. This new perspective raises new moral issues.

(iii) **Expanding moral concerns:** Some people argue that animals, trees and even rocks have moral and legal rights, and that it is a natural extension of civilisation to begin including the environment in ethics. These expanded concerns lead to need for new ethics.

Our effects on the environment today have consequences for the future, and therefore, discussions of environmental ethics also involve the rights of future generations. The arguments for and against various principles in environmental ethics are made more complex because of conflicting values. The resolution of the resulting conflict requires that we recognise differing values and that we have a basic scientific knowledge about the environment as well as the ability to clearly formulate a logical argument.

1.5 INTERNATIONAL ENVIRONMENTAL MOVEMENT

Environmental impacts of developmental activities are being sounded since the early 1960s. One of the first and most influential warnings was *Silent Spring,* a book about the use of pesticides, written by Rachel Carson and published in 1962. In the mid-1960s, some scientists were sounding warnings about the possibility of climate change due to increased carbon dioxide levels in the atmosphere resulting from burning of fossil fuels. The same period witnessed early environmental conferences like the two British "Countryside in 1970" conferences of 1963 and 1965, and the US White House conference on "Natural Beauty" of 1965.

In the United States of America (USA), this climate of thought and innovative legislators produced a law that represented a landmark in environmental management not only in the United States but worldwide – the National Environmental Policy Act (NEPA), 1969. The NEPA required "in agency planning and decision making an integrated interdisciplinary use of the natural and social services and the environmental design arts". It also required agencies to take account of non-quantified values, and to use ecological information and concepts in planning and decision-making affecting the quality of the environment.

"Spaceship Earth" arrived in the 1970's with concern growing, some warnings, and a piece of new style landmark legislation to be added to the many pieces of air and water pollution control law that existed already around the globe. The 1970's was an era of response and action. The most influential responses were the United Nations Conferences on the Environment (1972), Population (1974) and Human Settlements (1976).

The first UN Conference on Environmnent was held in June 1972 in Stockholm, Sweden. As a result, the United Nations Environment Programme (UNEP) was set up together with a fund to finance major projects. A 26-point declaration of environmental principles was adopted, calling for commitments by countries to deal with environmental problems of international significance. Approval was given to the Earth-Watch Program—a co-ordinated plan to use and expand existing monitoring systems to measure pollution levels around the world and their effects on climate. A convention to control ocean dumping of shore-generated wastes was endorsed. The International Whaling Commission was urged to adopt a ten-year moratorium on commercial whaling. Other proposals for conservation conventions were endorsed (e.g., the World Heritage Trust to protect areas of unique natural, historical or cultural value that are part of the heritage of all humankind, and a convention to protect endangered species of plants and animals, etc.).

Three months prior to the UN Conference a report on phase one of the Club of Rome's project on the *Predicament of Mankind* was published. *The Limits to Growth* described the work on a global model of the five factors that determine and limit growth – population, agricultural production, natural resources, industrial production and pollution. The Club of Rome first came together in 1968 out of the concern of thirty individuals from ten countries to discuss the present and future predicament of humanity and to foster understanding of the interdependent components that make up the global system.

Another conference, less visible than that of the United Nations, but like *Limits to Growth*, something as a pointer to future, occurred also in 1972. This was the Second World Conference on National Parks and Protected Areas, held under the aegis of the World Conservation Union (IUCN). IUCN was established in 1948, and its work and influence would become of increasing importance, as biodiversity and the still accelerating forces of population growth, technology and pollution reduced the gene stocks of wild nature. (In 1972 world population was 3.7 billion.) At this time the now well-known conflicts and concerns about the loss of tropical forests were yet to come. At this time also, conservation and development were seen as opposing, rather than complementary objectives as inferred by UNCED.

In parallel with, and following the Stockholm Conference, there was a surge of policy-making and institution building in the developed world. In the decade of the 1970's, many governments introduced environmental legislation, established agencies with environmental responsibilities, or grafted these onto existing departments. Others moved more cautiously, with 'pilot' policies that could be elaborated after experience had been gained. Environmental impact assessment (EIA), the discipline that could steer decision-making towards allowance for environmental factors was introduced in some countries. Often regarded with suspicion as an unnecessary impediment to legitimate developmental objectives and progress, EIA took a decade to be acknowledged as a tool that could actually produce projects superior in both quality and value.

In 1971, the World Bank established an environmental section to promote environmental reconnaissance of hydro-projects in recognition of one of the fundamentals of environmental management. The principle that knowledge of the environmental context of development proposals should be a base within which planning proceeds with constant reference to potential impacts. The major negative impacts of some hydro-projects had already included spread of water-borne disease. Loss of biodiversity due to destruction of habitat and wildlife, water quality effects on reservoirs and downstream locations, induced seismicity and decline of productive ecosystems. One of the objectives of the World Bank initiative was *capacity building*. Bank agents may have carried out the initial reconnaissance but borrowers were aided and encouraged towards developing competence and setting up of in-house environmental units.

Development of economic growth, technological expansion and exponential population increase continued and problems of global scale surfaced. By 1980, the handling and disposal of hazardous wastes was recognised as a national and global issue. Through the 1970s and 1980s, the growth of chemical manufacturing in developing countries exceeded that of the developed world. Samples showed that developing world populations ranked high in exposure to toxic chemicals, blood lead levels and DDT contamination of human milk. Effective national regulation and control programmes existed in developed countries, but not elsewhere. The discovery of grossly contaminated sites (a danger to water supplies, food chains, and human health generally) in the developed world could only raise questions, but not answers, about the developing world where disposal is haphazard, and neither disposal facilities nor national control frameworks exist.

In I970, most air pollution legislation treated air pollution as a local phenomenon. By 1980, it had become global and was affecting the forests and lakes of Europe and North America. The major cause was the burning of fossil fuels. Global emissions of sulphur dioxides grew by an estimated 470%, and emissions of carbon dioxide tenfold, in the 20th century. Nitrous oxide emissions in the USA have increased nine times over the same period. Given suitable atmospheric conditions, sulphur dioxides and nitrous oxides can be transported long distances and transformed into acids. It was realised that acid rain affected not only lakes and streams but also crops and vegetation.

At ground level, ozone is also damaging to crops and vegetation. But in the upper atmosphere ozone acts as a filter, absorbing harmful wavelengths of ultra-violet radiation. In 1985, it was confirmed that the ozone shield over the Antarctic was thinning, and a 'hole' had developed, and an unprecedented, international action followed for a world action plan. Its main components were global monitoring, estimating the impact of changes in the ozone layer on radiation, skin cancer, ecosystems and regional climate, and collecting data on production and emissions. While the world action plan was developing, an international convention was being hammered out. The Montreal Protocol, providing a framework for action by each country, was agreed in 1987. Officials from most of the CFC producing/using countries agreed to a 50% reduction by 1999. But, new scientific evidence indicated that the situation was more serious than had been thought. The Helsinki Declaration of 1989 stated the intention of 80 countries to completely phase out CFCs by the year 2000.

Another, and greater, global concern was also growing as the ozone saga unfolded. The concerns about increasing CO_2 levels, expressed by a few scientists in 1960s were confirmed in the 1980s by a large proportion of the scientific community. Data showed increasing concentrations, not only of CO_2 but also nitrogen dioxide (NO_2), methane (CH_4) and specific chlorofluorocarbons. Their addition to the atmosphere permits it to absorb more of the infrared radiation emitted from the earth.

The understanding and application of the concept that environment and development are integrated, complementary and indivisible, and are to be thought of as one, is fundamental to environmental management. This new message was given in 1980 in the World Conservation Strategy, published by the World Conservation Union (IUCN), the United Nations Environment Programme (UNEP) and the World Wide Fund for nature (WWF). Its successor document, *Caring for the Earth,* was published in 1991. This strategy, so clearly complement to *Our Common Future,* sets out the rationale and the strategic actions needed to achieve a sustainable future.

The original World Conservation Strategy of 1980 introduced an important innovation in environmental management – the national level conservation strategy, subsequently prepared by over 50 countries. These identify the country's most urgent environmental needs, assist decision-makers in determining priorities, allocate resources and build institutional capacity to handle complex environmental issues. Preparation includes fundamental re-examination of laws, policies and institution. Similar to these national conservation strategies were national environmental action plans sponsored by the World Bank.

In the early 1980s, the then Minister of the Environment of the government of Norway, Dr. Gro Harlem Brundtland worried by the growing incompatibility between human development and the possibility for nature to withstand the resulting ecological stress, undertook to launch a global investigation of this incompatibility. With the assistance of a former Vice President of the Republic of Sudan, she managed to assemble a team of 26 concerned persons of high calibre into what was named the World Commission on Environment and Development (WCED). The WCED project was given the green light—but not a full endorsement—by the United Nations and succeeded in obtaining support in kind and cash from a good number of countries. There were some noticeable exceptions to the supporting countries, such as the USA, Japan, France, the United Kingdom and Germany.

Nevertheless, the WCED had sufficient means to commission a wide range of necessary surveys and studies by competent and well known scientists and professionals, access the statistics of many countries and all international organizations, and undertake a series of hearings in a sample of countries around the world. Its exploration of the issues ended sometime in 1986. With the help of some countries (e.g., Canada), the WCED undertook to prepare a summary of its findings, with a synthesis, conclusions and recommendations. The result was a paperback report, under the title *Our Common Future*, published in April 1987 by Oxford University Press, gradually translated into a large number of languages and distributed widely around the world. The report was officially presented to the General Assembly of the United Nations in November 1987, and enthusiastically approved by most, if not all members.

At the core of the recommendations made by the Commission was the goal of *sustainable development*, which was defined as development that meets the needs of the present without compromising the ability of future generations to meet their own needs.

Our Common Future was a precursor, though not the only one, to United Nation's Conference on Environment and Development (UNCED) of June 1992 (popularly known as the Rio Earth Summit). Both the report and the Conference link environment and development, bringing together within the concept of systems and interdependence, environment and development objectives once considered to be in opposition. And, in 1992, *Agenda 21*, a global plan of action for sustainable development was adopted. In addition to adopting *Agenda 21*, the assembled leaders from over 100 countries signed the *Framework Convention on Climate Change* and the *Convention on Biological Diversity* and endorsed the *Rio Declaration* and the *Forest Principles*.

The Rio Declaration represents a set of 27 agreed principles aimed at the objective of a new and suitable global partnership, international agreements that respect the integrity of the global environmental and developmental system, and recognition of the integral and interdependent nature of earth, our home.

The outcomes of the discussion of *Agenda 21* occupy more than 450 pages. *Agenda 21* is the operational plan for moving humankind into the age of sustainability. It should be read and given profound consideration by all people, specially all technologists. Its implementation will demand commitment and the capacity of all nations, people and individuals.

More than half the sections of *Agenda 21* are of direct or indirect relevance to science and technology, e.g., those dealing with the protection of human health, human settlement, integration of environment and development in decision-making, protection of the atmosphere, integrated approaches to the planning and management of land resources, combating desertification and approaches to planning and management of land resources, combating desertification and drought, sustainable mountain development, conservation of biological diversity, environmentally sound management of biotechnology, toxic chemicals, hazardous wastes, solid wastes, radioactive wastes, the scientific and technological community, transfer of environmentally sound technology, education, national mechanisms, institutional arrangements and information.

The Commission on Sustainable Development (CSD) was created in December 1992 to ensure an effective follow-up of UNCED; to monitor and report on implementation of the Earth Summit agreements at the local, national, regional and international levels. The CSD is a functional commission of the UN Economic and Social Council (ECOSOC), with 53 members. It was agreed that a five-year review of Earth Summit progress would be made in 1997 by the United Nations General Assembly meeting in special session: Earth Summit + 5. The Special Session of the General Assembly held in June 1997 adopted a comprehensive document entitled *Programme for the Further Implementation of Agenda 21* prepared by CSD. It also adopted the programme of work of the Commission for 1998-2002.

Rio + 10, the Earth Summit + 10, the Johannesburg Summit or formally, *The World Summit on Sustainable Development* took place in Johannesburg, South Africa, 2-11 September, 2002. Thousands participated at the Johannesburg Summit: heads of State and Government, national delegates and representatives from business and industry, children and youth, farmers, indigenous people, local authorities, non-governmental organisations, scientific and technological communities, women and workers and trade unions.

According to General Assembly resolution 55/199 of 20th December 2000, the review would focus on the identification of accomplishments and areas where further efforts are needed to implement *Agenda 21* and other outcomes of the UNCED. At the national level, Governments are encouraged to prepare national assessments on the progress achieved and challenges encountered. At the regional level, the UN Department of Economic and Social Affairs (UNDESA) has sponsored round tables of eminent persons and UN regional commissions facilitated regional ministerial preparatory meetings. At the global level, the *UN Secretariat for the Summit* organised three substantive meetings in the first half of the year 2002 before convening the Summit in September.

Against the backdrop of the discussions we have had so far, we will next focus on the major environmental concerns in India.

1.6 ENVIRONMENTAL CONCERNS IN INDIA

We can categorise environmental concerns in India as those arising from negative effects of the very process of development; and conditions of poverty and under-development. The major problems, which encompass the area of EM are:

- Burgeoning population and its impact on life support systems, which negate the positive impacts of the developmental programmes.
- Rapid growth in the population of domesticated animals over the past few decades accompanied by a loss of area under grassland and pastures. Around 3.5% of the geographical area is under grasslands, while the domesticated animal population numbers nearly 500 million (1991).
- Out of a total area of about 329 million hectares, 175 million hectares of land require special treatment to restore them for productive and profitable use. Land degradation is caused by water and wind erosion (159 million ha), salinity and alkalinity (eight million ha), and river action and other factors (seven million ha).

- The forest wealth is dwindling due to overgrazing and over-exploitation both for commercial and household needs, encroachments, unsustainable practices including certain practices of shifting cultivation, and developmental activities such as roads, buildings, irrigation and power projects.
- The recorded forest cover in the country is about 75 million hectares, which is 19.5% of the total geographical area against the national goal of 33% in the plains and 66% for hilly regions. Even within this area, a meagre 11% constitutes forests with 40% or more of crown cover. The annual rate of loss of the forest cover is 47,500 hectares.
- The loss of habitat is leading to the extinction of plant, animal and microbial species. Over 1500 plant and animal species are in the endangered category.
- The wetlands of India, which are rich in aquatic and bird life, providing food and shelter as also the breeding and spawning ground for the marine and freshwater fishes, are facing problems of pollution and overexploitation.
- The major rivers of the country are facing problems of pollution and siltation. The coastline is under stress and coastal areas have been severely damaged due to indiscriminate construction, waste disposal near the waterline and aquaculture. Coastal vegetation including mangroves and sea grasses is facing extinction. The mountain ecosystems are under threat of serious degradation.
- Extensive deforestation leading to the erosion of valuable topsoil is threatening the livelihood and security of millions of hill people. The quality of groundwater is being affected due to chemical pollution and due to the ingress of seawater in coastal areas.
- Absence of an integrated land and water use policy for the country has resulted in a heavy toll of basic natural assets. Coral reef ecosystems are adversely affected by indiscriminate exploitation of corals for production of lime, recreational use and for ornamental trade. Island ecosystems are subjected to pressures of various forms, including migration of people from the mainland.
- Pollution arising from toxic wastes and non-biodegradable consumer articles is on the rise. A large number of industries and other development projects are sited close to heavily populated urban centres, leading congestion and pollution, as also the diversion of population and natural resources from the rural areas.
- Mechanism to indigenously arrive at a reliable estimate of total greenhouse gas emissions in the country, among various sources such as agriculture, animals, energy production and consumption, forestry and land-use change, waste management, etc., is inadequate. More such reliable data, which is indigenously arrived at, is essential for negotiating international law, treaties, protocols and conventions on environment-related problems where developing countries are unduly pressurised.

It is often difficult to clearly delineate the causes and consequences of environmental degradation in terms of simple cause-effect relationships. The causes and effects are often interwoven in complex webs of socio-economic, technological and implementation factors.

The liberalisation of the economy in India in the recent past is expected to promote consumerism by throwing a wide array of products with wide options to the consumers. This, in turn, is expected to put pressure on the natural resources in the form of raw materials, and therefore, may deplete the natural resources and cause irreversible damages. Several plant and animal species may become extinct and the non-renewable resources may be exhausted soon, depriving the future generations of their benefits. This has been amply witnessed in the past in several western countries where industrial revolution depleted the natural resources for generations to come. Today, these countries are looking towards the developing countries for raw material and natural resources. India should learn a lesson from the past experiences of industrial development within India and outside of it.

LEARNING ACTIVITY 1.6

In your view, is it possible to integrate environmental and developmental concerns? Give at least three examples to support your answer.

Note:

a) Write your answer in the space given below.

b) Check your answer with the one given at the end of this book.

SUMMARY

In this Unit, we provided you with an overview of environmental management and pointed out that it is an extensive but loosely defined field. We then outlined the need for ecosystem approach in environmental management, the types of management processes and tools, and the responsibilities of professionals. The understanding of the environmental concerns and initiatives, we said, has led to the evolution of various approaches including the ecosystem approach. We also discussed some of the issues relating to environmental concerns and ethics in the management of environment. Finally, after providing an account of the global environmental movement, we presented the concerns in India.

REFERENCES

Adam, P. 1990. Saltmarsh Ecology. Cambridge University Press, Cambridge

Barbier E B, 1987. The concept of sustainable development, Environmental Conservation, 14 (2): 101-110.

Botkin,D. B. and Keller E. A, 1995. Environmental Science: Earth as a Living Planet, John Wiley, N. Y.

Dorney, R. S, 1989. The Professional Practice of Environmental Management. Springer Publication, Heidelberg.

Harrison, P. 1990. Beyond the Blame Game: population environment links, Populi 17(3): 14-21.

Jolly, V, 1978. The Concept of Environmental Management Development, Foram, 8 (2), 13-26.

Klee, G. A. 1980. World System of Traditional Resource Management, J Wiley and Sons, N.Y.

MOEF, 1999. National Policy and Macro Level Nation Strategy on Biodiversity, The Ministry of Environment and Forest, Government of India, New Delhi.

Mukherjee, 2002. Ecosystem Concepts, http://www.scribd.com/doc/4652798/ECOSYSTEM-CONCEPTS

Selin S and Chavez D, 1995. Developing a collaborative model for environmental planning and management. Environmental Management 192 (1995), pp. 189-195.

The Ecologist, 1993. Whose Common Future? reclaiming the commons, London: Earthscan, p.107

Turnham D, 1991. Multilateral development banks and environmental management Public Administration and Development, Volume 11, Issue 4, Pp. 363-379

Vogler J and Imber M,1995. The Environment and International Relations, London: Routledge, 1996

WCED, 1987. Our Common Future, United Nations, World Commission on Environment and Development, Oxford University Press, London.

SUGGESTED READINGS

Angell, D.J.R., Comer, J.D. and Wilkinson, M.L.N. (ed). 1990. *Sustaining Earth: Response to the Environmental Threat*. Macmillan, Basingstoke.

Bergesen, H.O. and Parmann, G.J. 1994. *Green Globe Year Book. 1994.* Oxford University Press, Oxford.

Berry, R.L. (editor). 1992. *Environmental Dilemmas, Ethics and Decisions.* Chapman and Hall; London.

Commonwealth of Australia, 1991. Ecologically Sustainable Development Working Groups, *Greenhouse Report,* Australian Government Publishing Service, Canberra.

Ellis, D. 1989. *Environments at Risk, Case Histories of Impact Assessment.* Springer-Verlag, Berlin.

IUCN, UNEP, WWF. 1991. *Caring for the Earth. a strategy for sustainable living.* IUCN (The World Conservation Union), UNEP (United Nations Environment Programme), WWF (World Wide Fund for Nature), Gland, Switzerland.

Hansen, P.E. and Jorgensen, P.E. 1991. *Introduction to Environmental Management.* Elsevier, Amsterdam.

Institution of Engineers, Australia. 1992. *Environmental Principles for Engineers.* Canberra.

McMichael, A.J. 1993. Planetary Overload: Global Environmental Change and Health of Human Species. Cambridge University Press.

Meadows, D.H., Meadows, D.L., Randers, J. and Behrens, W.W. 1972. *The Limits of Growth.* Earth Island Ltd, London.

Meadows. D.H., Meadows, D.L. and Randers, J. 1992. *Beyond the Limits; Global Collapse or Sustainable Future.* Earthscan Publications Ltd, London.

Moore, James W. 1986. *The Changing Environment.* Springer-Verlag, New York.

Nash, R. 1989. *The Rights of Nature. A History of Environmental Ethics.* University of Wisconsin (There are also other publishers.)

Nath, B., Hens. L., Compton, P. and Devuyst, D. 1993. *Environmental Management, Volume 1. Compartmental Approach.* VUB University Press, Brussels.

Nath, B., Hens, L., Compton, P. and Devuyst, D. 1993. *Environmental Management, Volume 2. Ecosystems Approach.* VUB University Press, Brussels.

Nath, B., Hens, L. and Devuyst, D. 1993. *Environmental Management, Volume 3. Instruments, Implementation.* VUB University Press, Brussels.

National Geographic Magazine. 1988. *Can Man Save this Fragile Earth?* Vol 174, No. 6. National Geographic Society, Washington, D.C.

OECD. 1994. *Environmental Indicators—OECD Core Set.* OECD, Paris.

Pearce, D., Barbier, E. and Markandya, A. 1990. *Sustainable Development, Economics and Environment in the Third World.* Earthscan, London.

Ryding, S-O. (Editor). 1994. *Environmental Management Handbook,* 2nd Edition. IDS Press. Amsterdam.

Scientific American Magazine. 1989. Special Issue: *Managing Planet Earth,* September, 1989.

World Bank. 1993. *The World Bank and the Environment, A Progress Report: Fiscal 1993.* World Bank; New York.

World Commission on Environment and Development. 1987. *Our Common Future.* Oxford University Press, Oxford.

UNIT 2

Policy and Legal Aspects of EM

STRUCTURE

OVERVIEW

In Unit 1, we introduced you to the fundamental principles and concepts of the environment and ecosystem. In this Unit (i.e., Unit 2), we will provide you with the background to the evolution of environmental policies and legislations in India. We will begin the Unit by explaining the international environmental policies in general. This will give us a context for our discussion of the policies in India. We will also discuss some of the salient features of environmental laws, the NGO movement and legislations, as they pertain to India.

LEARNING OBJECTIVES

After completing this Unit, you should be able to:

- discuss the evolution of environmental policies and laws;
- explain the implications of international policies and laws for India.

2.1 INTRODUCTION TO ENVIRONMENTAL POLICIES

Environmental policies may be either enacted as laws by governing bodies or created and enforced by government agencies. They may originate from local, national or foreign governments, and address an array of issues including (but not limited to) air or water quality, fossil fuel extraction, energy conservation, habitat protection or restoration, pesticide use, storage/disposal of hazardous materials, recycling and trafficking in endangered species.

An environmental policy being interdisciplinary in nature draws together technology, economics, and natural and social sciences. In order to develop sustainable policies, therefore, it is necessary to have sound knowledge of the actual and potential environmental impacts of certain activities and some knowledge of the technical characteristics, economic costs, social acceptability and possible side effects of alternative policy options.

The quality of the environment has both a direct and an indirect effect on the standard of living. This does not mean that environmental degradation is simply a by-product of economic activities, it is also the consequence of the priorities set by states in their economic policies. These policies generally aim at stimulating production and, as a consequence, tend to ignore their implications for the environment. Past experience, however, shows that economic policies may actually have more impact on the quality of the environment than those policies explicitly designed to protect the environment. We will discuss this, next.

2.1.1 Economics and Environmental Policies

One sign of a sustainable economy is when the costs of environment and health caused by economic growth have been added to consumer prices and when economic policy instruments support sustainable development. Environmental policies should supplement economic instruments.

Environmental policies involve certain measures aimed at achieving a sound environment. They are usually developed in the context of public policy, based on economic theory, which focuses more on the level of costs and benefits associated with the implementation of environmental policies than on the quality of the environment. When governments propose and subsequently implement strict standards, sectors that pollute the environment will have to take measures, and this cannot be achieved without incurring extra costs. Polluting industries are, therefore, often keen to highlight the likely costs they have to incur due to the proposed environmental measures.

In other words, the immediate benefits resulting from environmental policies are extremely difficult to assess. As a consequence, the costs of environmental measures are often paid more attention than the benefits resulting from the implementation of the policy.

The definition of the property rights of natural resources plays a vital role in the distributional effects of environmental policies. The implementation of strict standards and regulations will effect a change in the definition of property rights. For example, industries polluting the rivers will be confronted with regulations that prevent them from, or reduce their opportunities for, using the rivers. However, throughout the process of formulating the regulations, polluting industries will try to influence and stifle the policies. (Note that we will discuss some of the industrial policies as they pertain to the environment in Subsection 2.1.2)

Let us now consider below a few examples of sectorial economic policies that influence the environmental policies directly or indirectly:

- **Agricultural sector:** Virtually the entire food cycle attracts huge direct or indirect subsidies, at a cost to taxpayers and consumers. These subsidies, more often than not, send farmers far more powerful signals than do the small grants, usually provided for soil and water conservation. They encourage farmers to occupy marginal land and to clear forests and woodlands, make excessive use of pesticides and fertilisers, and use underground and surface waters in irrigation indiscriminately. We will discuss the policies of the agricultural sector in detail in Subsection 2.1.3.
- **Forestry sector:** The pressures on forests throughout the world vary greatly in both developed and developing countries, which are reinforced by government policies. The logging and forestry industry attracts a variety of direct and indirect subsidies. The perverse incentives that encourage the overharvesting of temperate as well as tropical forests also mark world-trade in forest products.
- **Transport sector:** This sector, especially motor vehicles, also benefits from economic policies that are ecologically perverse. Fuel taxes in many jurisdictions, for example, still fail to distinguish between the environmental effects of different types of fuel (i.e., petrol or diesel, leaded or unleaded). The tax and tariff structure, and direct and indirect subsidies, encourage heavier and more energy-intensive vehicles and road freight, as opposed to rail transport in many countries. In addition, in some countries, private vehicle expenses can be deducted from taxable income.
- **Energy sector:** The major obstacle to energy efficiency is the existing framework of incentives for energy exploration, development and consumption. These incentives underwrite coal, oil and gas, ignore the costs of air, land and water pollution and seem to favour inefficiency and waste. While industrialised

countries have been spending billions to distort the market and consumer prices in ways that actively promote acid rain and global warming, they have been spending only a few million on measures to promote energy efficiency. As long as pollution problems are mainly national, there is a need for a strong national authority. However, environmental problems are becoming increasingly international or global. This complicates the environmental policies considerably. On the one hand, international cooperation in the fighting of environmental problems is absolutely necessary while, on the other hand, each country has different economic interests. Furthermore, polluting sectors are not evenly distributed among countries.

2.1.2 Industries and Environmental Policies

Industries are a measure of a country's economic growth. Consequently, countries have a tendency to protect their polluting industries, in particular when they are relatively important economically. However, the growing interest in environmental management has fuelled certain industries to adopt policies that are economically feasible and which helps curb environmental degradation.

Various factors drive the development of a managed approach to environmental performance. These include the following:

- The need to meet increasingly stricter environmental regulations.
- Stakeholder pressure (e.g., pressure from shareholders, insurers and investors).
- Supply-chain pressure from customers.
- Historically poor relations with regulatory bodies and local communities.

Many industries have established environmental management systems (EMS) to tackle activities, which either pose a serious threat to ecosystems in the event of accidents or involve significant expenditure because of the costs associated with raw material use and/or waste disposal. An EMS is "the part of an overall management system that includes organisational structure, planning activities, responsibilities, practices, procedures, processes and resources for developing, implementing, achieving reviewing and maintaining the environmental policy" (BSI, 1996; see *http://www.env.hie.co.uk/welcome.asp.LocID_envdev.htm* for details). An EMS aims to help organisations achieve sound environmental performance by identifying key activities which impact, already or potentially, on the environment and by putting in place management controls to ensure that the organisation continues to meet its legal and policy requirements to deal with these impacts.

Traditionally, most of the work on EMS has been done by those industrial sectors with the greatest potential to affect the environment, e.g., the chemical, waste management and oil refining sectors. In addition to this basic need to control pollution, some companies have identified EMS as an effective way of improving productivity through waste and resource minimisation initiatives and a mechanism for increasing sales as customers turn on to environmentally safer products.

Arguably, implementing effective EMS can be a useful de-regulatory tool for business. For example, by reducing the use of certain hazardous substances, some organisations in India have achieved major improvements in their air emissions and, as a result, they came out of the strict controls imposed on them by the Environmental Protection Act (EPA) 1990. In addition, they have gained a better working environment for their employees and eliminated a difficult raw material storage hazard.

Close on the heel of industrial policy is the agricultural policy that can be adopted to prevent the deleterious effect of agricultural activities on the environment. Let us discuss this issue in Subsection 2.1.3.

2.1.3 Agriculture and Environmental Policies

Agriculture has a major impact on the environment, especially on land, water and biodiversity. Over the last 10-15 years, the environmental performance of agriculture has been mixed. For example, nitrogen and pesticide loading in water remain relatively high and risks of soil erosion and water resource depletion persist in many regions and countries. In recent years, however, there have been improvements in wildlife habitats, landscapes and sinks for greenhouse gases provided by agriculture, but the most significant progress has occurred where environmental pressures have been greatest.

The main environmental impacts of agriculture may be characterised through the beneficial or harmful contribution of agricultural activities to:

- soil quality (e.g., erosion, nutrient supply, moisture balance, salinity, etc.);
- land quality (e.g., ecological management of agricultural land);
- water quality (e.g., nutrient, pesticide and sediment runoff and leaching, salinity);
- water quantity (e.g., irrigation consumption, use efficiency, water retention capacity, flood prevention, etc.);
- air quality (e.g., emissions of dust, odours, ammonia and greenhouse gas, absorption of carbon dioxide, etc.);
- biodiversity (e.g., farm and indigenous animal and plant diversity);
- wildlife and semi-natural habitats (e.g., diversity of animal and plant habitats associated with farming);
- rural landscape (e.g., environmental features of areas shaped by farming, including those associated with historic buildings and landmarks, etc.).

Agricultural policies in India provide substantial farm support, often linked to commodity production affecting resource use, farming practices and environmental performance. Reconciling food production and environmental goals, however, is a challenge. But, reconciling them implies that the rights and responsibilities of farmers regarding farm practices need to be clearly defined and applied, and thus the situations under which they are entitled to remuneration or obliged to pay (polluter-pays-principle, or PPP). Defining who pays and who is paid for the desired level of environmental performance has important implications for the distribution of income and wealth.

Role of Agro-environmental Policies

When private and public mechanisms designed to facilitate the improvement and diffusion of appropriate farming practices and market forces are not enough to ensure the supply to meet the demand of environmental services, specific agriculture-environmental measures at the farm level may be needed. Such measures may be necessary to reduce the environmental harm, or enhance the environmental benefits of farming activities. When designing and implementing such policy measures, a number of general policy principles should be taken into account in the choice of the type of policy incentive or disincentive – payment or tax.

The PPP applies to reducing environmental harm for which farmers, as any other polluter, should be accountable. However, the PPP guiding principles recognise the possibility of different property rights and reference levels among countries, with the possibility of offering transitional financial incentives to encourage farmers to adopt appropriate production practices for improving their environmental performance through reducing environmental harm. It includes the case of transitional financial assistance provided to stimulate the development of new pollution control technologies and abatement equipment to achieve a better environmental performance through improved production practices.

General Policy Principles

When markets do not exist to allocate costs and benefits of agriculture-environmental impacts and outputs, policy action may be needed to account for the costs of not respecting environmental targets and to ensure the provision of environmental benefits. When designing and implementing policy measures, the environmental problem needs to be clearly defined and the following principles for policy design need to be kept in mind:

- The necessary condition for a welfare gain from implementing an agriculture-environmental policy measure is that the resulting environmental benefits exceed the costs associated with the policy. These costs include those due to a reduction in outputs, associated with more environmentally friendly technologies and practices and the transaction (administrative) costs of policy implementation and enforcement.

- When farmers and other economic agents provide a specific environmental service, the level of benefit should be clearly specified and efforts made to ensure that the most efficient operator is the provider.
- When a specific environmental outcome is jointly the result of agricultural output, a wide range of policy options and approaches may achieve its provision by an individual farmer that either provides positive incentives (through, for example, a payment) or negative incentives (e.g., a tax). If incentives were set correctly, it would be in the individual farmer's interest to achieve the outcome and receive the incentive payment, or achieve the outcome and avoid paying the tax.
- The effectiveness of either a tax or a payment depends not only on whether it correctly confronts the farmer with the opportunity costs of not respecting environmental requirements, but also on the degree to which the associated obligations can be enforced and tailored to local environmental circumstances and demands. The more the payment or tax is tailored to specific circumstances, the larger the need for monitoring, the lower the probability of individual control, and the higher the transaction costs.

In addition to what we have discussed, an environmental policy that conserves the ecosystem is a sagacious attempt to manage the environment. We will explain this in Subsection 2.1.4.

2.1.4 Ecosystem and Environmental Policies

The objective of ecosystem conservation and management as stipulated in the Forest Policy (MNRT, 1998a) is to ensure ecosystem stability through conservation of forest biodiversity, water catchments and soil fertility. Forest biodiversity is faced with problems of encroachment, shifting cultivation, wild fires, lack of systematic forest management and inadequate infrastructure and staff to prevent excessive resource use.

Watershed management and soil conservation face similar problems as biodiversity conservation due to increase in population pressure and inefficient forest management and protection in watershed areas. Poor management and protection have resulted in deterioration of watershed areas causing water shortages. Planting of inappropriate species in watershed areas, illegal logging and inappropriate logging methods have further reduced the quantity and quality of water and are the cause of peak floods, droughts and sedimentation in rivers. Erosion due to cultivation on the riverbanks outside forest reserves is also a major concern in watershed management.

Encroachment, wildfires, illegal logging and poaching are also the major factors contributing to the deterioration of wildlife populations in natural forests. Inadequacy of baseline data on types of wildlife species, their habitats and the lack of incorporating wildlife management in forest management plans, are the major reasons hampering wildlife conservation in forest reserves. Also, coordination among the government institutions involved in wildlife and forest management is very poor.

In order to minimise the damage to the ecosystem due to human activities, the Forest Policy stipulates environmental impact assessment (EIA) before implementing developmental projects, which convert forest land to other land uses such as timber industries, mining, road construction, agriculture, dams, settlements, shrimp farming and tourism. The scope and guidelines of an EIA is to be prepared in collaboration with other sectors and stakeholders.

Specifically, in order to achieve ecosystem conservation and management in the areas of biodiversity, watershed management and soil conservation, and wildlife, we need to consider:

- New forest reserves for biodiversity conservation in areas of high biodiversity value. Forest reserves with protection objectives of national strategic importance may be declared as nature reserves.
- Biodiversity conservation and management is to be included in the management plans for all protection of forests. Involvement of local communities and other stakeholders in conservation and management is to be encouraged through joint management agreements.
- Biodiversity research and information dissemination should be strengthened in order to improve biodiversity conservation and management.
- Biodiversity conservation must be incorporated in the management regimes of natural production forests and plantations. Management plans must incorporate biodiversity conservation and management guidelines. This minimises the replacement of natural forests by exotic plantations.
- Watershed management and soil conservation should be included in the management plans for all protection and production forests. Involvement of local communities and other stakeholders in watershed management and soil conservation will be encouraged through joint management agreements.
- Research and information dissemination must be strengthened in order to improve watershed management and soil conservation.
- EIA is required for the investments, which convert forestland to other land use or may cause potential damage to the forest environment.

Having looked at some of the sectoral policies that influence environment policies, let us now discuss the instruments used in implementing these policies.

2.1.5 Environmental Policy Instruments (EPI)

Environmental policy instruments (EPI) usually refer to official actions taken to curb and remove the negative environmental impacts caused by society. The methods, laws, administration and decisions relating to these actions are collectively termed environmental policy. An EPI is divided into economic, information and legal measures. Indicators of sustainable development frequently measure the status of development or pressures directed at it. Indicators of environmental policy instruments reflect a society's reaction, and the steps taken to make sustainable development possible.

A number of different typologies are used to classify EPI. One of the most commonly used typologies divides an EPI into the following three categories:

(i) Regulatory instruments that mandate specific behaviour.
(ii) Market-based instruments that act as incentives for particular activities.
(iii) Information-based instruments that seek to change behaviour through the provision of information.

We will discuss each of these, next. But, let us first note that governments may establish formal cleaner production strategies or programmes to act as a framework for the coordinated implementation of subsequent and more specific policy instruments. Cleaner production strategies may take one or a combination of the following shapes:

- **Product bans:** The imposition of a ban or defined phase-out schedule for a particular product or substance is an authoritarian means of promoting cleaner production. This may be implemented through application of the product choice or substitution principle.
- **Extended producer responsibility (EPR):** EPR aims at making environmental improvements throughout the life cycle of a product by making the manufacturer responsible for various aspects of the product's life cycle. In particular, this could include the take-back recycling and final disposal of the product.
- **Cleaner production audits:** As part of their permitting requirements, it is mandatory for production industries to carry out cleaner production audits of their plants and to implement findings as long as they do not harm the environment.
- **Mandatory EMS and reporting:** In terms of integrated permit conditions, it is mandatory for production industries to implement a structured environmental management system and make public information on their environmental performance.
- **Financial and technical incentives:** Governments may stimulate cleaner production measures by providing grants, loans and favourable tax regimes,

and/or by supplying targeted technical assistance to relevant industrial enterprises.

Regulatory Instruments

Since the inception of environmental policy, the predominant strategy for pollution control has generally been through the use of regulatory instruments. Usually, a public authority sets standards, and then inspects, monitors and enforces compliance to these standards, punishing transgressions with formal legal sanction. These regulations may, for example, specify an environmental goal such as the reduction of carbon dioxide emissions by a specified date. They may also mandate the use of a particular technology or process. Such an approach gives the regulator the maximum authority to control where and how resources will be allocated to achieve environmental objectives. Also, this provides the regulator with a reasonable degree of predictability as to how much the pollution levels will be reduced.

Specified and Negotiated Compliance

The specified compliance approach imposes the precise and specific demands on the regulated communities without any scope for bargaining and exceptions. However, this approach has some significant drawbacks in that the regulated community may tend to become alienated from the authorities and united in its opposition to the rules and regulations.

The negotiated compliance approach, by contrast, adopts a more cooperative approach between the regulators and the regulated in setting and enforcing standards. This shared responsibility between the government and industry enhances the likelihood of a more open exchange of information between the parties and allows greater flexibility regarding the means of meeting the standard. Moreover, a number of countries have started to develop regulations where attainment of certain targets (e.g., recycling targets) is required while concrete means of achieving such targets are left in the hands of industries (non-prescriptive regulations). This, in turn, may increase the economic efficiency of the regulation and may be conducive to the adoption of innovative, preventive approaches.

Indeed, with the growing appreciation of the limits of conventional policy instruments, many governments are encouraging the adoption of self-regulatory and co-regulatory policy instruments for promoting cleaner production.

Market-based Instruments

Market-based instruments generally seek to address the market failure of environmental externalities either by incorporating the external cost of a firm's polluting activities into the firm's private cost (for example, through taxation), or by creating property rights and facilitating the establishment of a proxy market (for example, by using tradable pollution permits).

Before introducing any new economic instruments, governments should identify and evaluate any economic incentives that may already be in operation, either explicitly or implicitly. These include, for example, the use of subsidies to make local industries more competitive. Many of these policies lead to artificially low prices for resources, such as energy and water, and as a result of which these resources may be overused, creating both pollution and shortages. Government assessments of such policies are, therefore, needed before other economic instruments are applied.

Taxes, fees and charges may be used to promote cleaner production practices by raising the costs of unwanted outputs or by providing incentives to promote more efficient use of natural resources. In some instances, it may be appropriate to use the revenues generated from these instruments to support cleaner production activities and thereby stimulating preventive approaches. A significant constraint against the more widespread adoption of market-based instruments, however, is that it is not always politically feasible to set taxes at a sufficiently high level to achieve desired environmental goals. Governments often face resistance, if taxation related to environment is taken merely as a means of increasing its revenues.

Governments may be able to avoid some obstacles by earmarking the corrected charges or shifting tax sources. In any case, the successful implementation of such instruments requires a system of monitoring, revenue collection and enforcement as well as measures to combat possible corruption.

Financial subsidies (e.g., low-interest loans, direct grants or preferential tax treatment) can be targeted to specific industries to stimulate technological development. Governments must, however, carefully examine how subsidies work to ensure that they are not misused resulting in environmentally counterproductive behaviour.

Information-based Strategies

In addition to creating an appropriate regulatory and financial framework for cleaner production, government may further stimulate the adoption of cleaner production practices through the use of informational measures. These may be used to provide the right incentive (e.g., through the public disclosure of a firm's environmental performance) as well as to build capacity within industry (e.g., through the publication and dissemination of relevant case studies).

A few examples of information-based strategies are given below:

- Promoting the adoption of targeted, high profile demonstration projects, to demonstrate the techniques and cost-saving opportunities associated with cleaner production.

- Encouraging educational institutions to incorporate preventive environmental management within their curricula, particularly within engineering and business courses.
- Requiring public disclosure of information on environmental performance by, for example, establishing a pollutant release and transfer register, stimulating greater voluntary corporate reporting and requiring the provision of information on specific materials.
- Initiating and/or supporting measures that address consumption such as eco-labelling schemes and environmental product declarations.
- Promoting the adoption of effective training initiatives.
- Issuing high profile awards for enterprises that have effectively implemented cleaner production.

  (www.uneptie.org/pc/cp/understanding_cp/cp_policies.htm).

Note that in some instances, an EPI is characterised by more than one of the above categories. Furthermore, it is important to note that different policy instruments are sometimes best used in conjunction with others.

LEARNING ACTIVITY 2.1

List any two policy instruments of cleaner production.

Note:

a) Write your answer in the space given below.
b) Check your answer with the one given at the end of this book.

Let us next discuss the evolution of environmental policies and programmes with particular reference to India.

2.2 ENVIRONMENTAL POLICIES AND PROGRAMMES IN INDIA

The year 1972 marks a watershed in the history of environmental management in India. Prior to 1972, different government ministries dealt with environmental concerns such as sewage disposal, sanitation and public health, and each pursued these objectives in the absence of a proper coordination system at the governmental or the intergovernmental level. During the twenty-fourth UN General Assembly of 1972, Mr. Pitamber Pant, a member of the Planning Commission of India prepared three reports about India. With the help of these reports, the impact of the population explosion on the natural environment and the existing state of environmental problems were examined.

The Planning Commission subsequently set up an expert committee to formulate long-term sectoral (including environment and forest) policies. It also noted that many environmental problems were continuing to cause serious concern, for example, the loss of top soil and vegetative cover, the degradation of forests, continuing pollution by toxic substances, careless industrial and agricultural practices and unplanned urban growth. It acknowledged that environmental degradation was seriously threatening the economic and social progress of the country.

The continuing decline in the quality of the environment spurred the Union, government and a few State governments to adopt stronger environmental policies, to enact fresh legislation and to create, reorganise and expand administrative agencies. Based on the recommendations of the Union Government in April 1990, the Government of India adopted a National Conservation Strategy (NCS) and Policy Statement on Environment and Development. The preamble to the NCS adopts the policy of sustainable development and declares the government's commitment to reorient policies and action in unison with the environmental perspective. The NCS proceeds to recognise the enormous dimensions of the environmental problems faced by India and declares strategies for action in various spheres such as agriculture, forestry, industrial development, mining and tourism.

In February 1992, the Union Government published its policy for the abatement of pollution. This statement declares the objective of the government to integrate environmental considerations into decision-making at all levels. To achieve this goal, the statement adopts some fundamental guiding principles: prevention of pollution at source, adoption of the best available technology, the polluter pays principle and public participation in decision-making.

Thus, the policies and legislation for environmental management in India has evolved over the years to meet the needs of forest conservation for sustainable development and environmental protection for improvement and management of ambient quality of air, water and soil.

In Subsections 2.2.1 and 2.2.2, we will discuss these and non-government organisations (NGOs), which were instrumental in stalling many industrial projects that would have otherwise had a deleterious effect on the environment.

2.2.1 Forest Conservation Activities

India is well known for its *aranya sanskriti* or forest culture. Forest dwelling communities lived in symbiosis with a large number of wild animals, hunting some and venerating others. With the advance of agriculture with the Iron Age, forests began their retreat to the hilly regions of the Himalayas and their lesser ranges, the central Indian plateaus and Western and Eastern *ghats*. The practices of indigenous forest tribes and people of the Northern and North-Eastern hills still demonstrate the close relationship that formerly existed among human beings, forest and other species.

The gradual decline of forests led to a number of seasonal, ritual and specific restrictions on forest-based activities. They were reinforced by religious and totemic proscriptions against harming certain types of trees and birds, snakes and animals. India's forest tribes still worship a number of trees and natural phenomena. For example, the annual hunt of the *Santhals* is preceded by an invocation to the deities of the forests and animals. Even the act of cutting the branches of certain trees must be accompanied with a verbal, personally communicated apology to the sacred tree.

The destruction of the last reserves of natural forests, however, has its roots in the British colonial policy towards the commercial exploitation of forests. With the introduction of the comprehensive Indian Forest Act in 1878, the colonial state radically redefined proprietary rights, imposing on the forest a system of management and control, the priorities of which sharply conflicted with earlier systems of local use and control. The clear felling of extensive natural forests and introduction of scientific management for their commercial exploitation marked an ecological, economic watershed in Indian forestry history.

The Forest Act of 1878 abolished traditional rights and customary practices governing forest management, and the Indian Forest Act of 1927 gave total control over forest resources to the State. This process continued in independent India, which retained the forest laws of the colonial government. The objective of maximising revenues from the sale of what was once common property resource was now done in the name of national interest and economic growth.

The Forest Policy of 1952, subsequently, emphasised the exploitation of forests primarily for producing valuable timber for industry and other vital purposes of national interest. The policy also strongly emphasised that proximity to forest resources did not ensure village communities rights over the property. The National

Commission on Agriculture, 1976, drew up a second policy in favour of commercial viability of forests and production of wood for industrial purposes.

These two policies led to a large reduction in the natural forest cover. This was mainly due to the very low concessional rates charged to industries for raw material. The major negative effects were the non-regeneration of forests due to unsustainable commercial exploitation, denial of peoples' legitimate rights to forest products and imbalance in ecologically sensitive regions leading to landslides, soil erosion and loss of biodiversity.

LEARNING ACTIVITY 2.2

Name some of the types of afforestation programmes in India.

Note:

a) Write your answer in the space given below.

b) Check your answer with the one given at the end of this book.

2.2.2 NGO Movements for Environmental Protection in India

The urgency of industrial development in the urban areas often ignored the needs of the environment, and the feeble voices of the local community were drowned in the publicity of large organisations. As a result, several committed individuals came forth and formed NGO to give voice to the feeble. Let us touch upon some key NGO movements in India.

Silent Valley Struggle

In the 1970s, led by *Kerala Sastra Sahitya Parishad*, students and a galaxy of intellectuals, academics, artists and naturalists conducted a nationwide movement against a hydroelectric project that would have destroyed part of the Silent Valley, a unique tract of 9000 hectares of forest in Western Ghats. In those times, none of the government departments/agencies was competent to deal with this kind of public opposition to developmental projects endangering the environment. Ultimately, decisions were taken at the level of the Prime Minister to stop the construction of the hydroelectric dam project. The discovery of nine previously unknown plant species and a new genus by the Botanical Survey of India in the Silent Valley reinforced the activists' determination to protect the forest.

Chipko Movement

In the 1970s, a disastrous flood occurred in the Alaknanda river in Garhwal. Local people were the first to link the disaster with the ecological change that had taken place in the watershed due to felling of forests. As a result, a movement was born in March 1973 in the tiny hamlet of Gopeshwar in the Garhwal region. Here, the villagers were denied access to ash wood for their agricultural implements. When the trees were allotted to a sports good manufacturing company, villagers rushed to protect the same trees from contractor's axe by hugging them. The method of protecting trees by hugging them spread like wildfire throughout the Garhwal region in the next few years. The movement, first interpreted as a fight for people's rights over scarce common property resources, quickly became a true ecological movement, in that it recognised the importance of forests in maintaining the fragile ecosystem. Garhwal villagers asserted that forests were the foundation of both their cultural and material life and, therefore, their birthright. Religious leaders supported the Chipko movement and Gandhian disciples spearheaded its manifestations. There were also folklore reflecting the spiritual and ecological foundation of the *Satyagraha*.

Chipco succeeded in changing the official strategy of forest exploitation to one responsive to the needs of people. By 1977, it became a full-fledged ecological movement, recognised at the highest level in the policy-making. A moratorium was placed on commercial green felling in the Himalayan region of Uttar Pradesh. Following this, Mr. Sunderlal Bahuguna and other satyagrahis conducted a 4780-km-long march from Garhwal to Kohima to spread the Chipco message. In Garhwal, the *Dashauli Gram Swarajyasangh* organised the largest voluntary afforestation programme in the country, planting millions of trees to restore the barren hillsides.

Public Interest Litigation for the Protection of the Taj Mahal

The Mathura Refinery is located in the vicinity of the Taj Mahal in Agra, despite a joint-Parliamentary Committee expressing its serious concerns about the location of the refinery within the Taj Trapezium, a 10,400 km^2 area. The Mathura Refinery and more than 2300 small and medium industries around it, including iron foundries, have continuously been emitting tonnes of pollutants such as sulphur dioxide and other poisonous gases, yellowing the Taj Mahal. Environmental Advocate, Mr. M.C. Mehta, filed a public interest litigation in the Supreme Court for the protection of the Taj from serious pollution. A bench of the Supreme Court has passed a number of orders directing the industries to install devices to control pollution or else face closure, the Mathura Refinery to minimise sulphur dioxide emissions and the Union Ministry of Environment and Forests to take suitable measures to save the Taj.

So far, we have been discussing the various environmental policies aimed at the protection of the environment. Let us now discuss the legal aspects of environmental management. Before we do so, we will first do Learning Activity 2.3.

LEARNING ACTIVITY 2.3

List five NGOs working for environmental protection.

Note:

a) Write your answer in the space given below.

b) Check your answer with the one given at the end of this book.

2.3 ENVIRONMENTAL LAWS AND LEGISLATIONS

Environmental law refers to rules and regulations governing human conduct likely to affect the environment. It reflects the legislative measures, and the administrative and judicial structures to protect the environment. However, it is difficult to define precisely the boundaries of environmental law in the same way as we define, say, the law of contract. Unlike the traditional legal subjects such as contract, which are well developed, environmental law is still in its infancy. Nevertheless, attention is now increasingly focused on the rationalisation and streamlining of existing measures rather than the development of substantial law.

Environmental law aids in:

- regulation of resource use;
- protection of the environment and biodiversity;
- mediation, conflict resolution and conciliation;
- formulation of stable, unambiguous undertakings and agreements.

Legislations have evolved in response to problems, so that there is often a delay between the need and the establishment of satisfactory law. Without effective legislation, resource use, pollution control, conservation and most fields of human activity are likely to fall into chaos and conflict. Law can encourage satisfactory performance, enable authorities to punish those who infringe environmental management legislation, confiscate faulty equipment or close a company. It may also be possible for employees, bystanders and product or service users to sue for damages, if they are harmed. Environmental laws, in essence, are indispensable instruments in curbing environmental degradation.

2.3.1 Principles of International Law

International law governs relations between States and has no direct effect on domestic law or individuals. It is often difficult to force a sovereign State to sign and then honour a treaty or similar agreement. International law, thus, depends a great deal on voluntary agreements by governments and international bodies.

The 1972 UN Conference on the Human Environment (held in Stockholm) is rightly regarded as the starting point of the development of modern-day international environmental law. The conference marked the worldwide recognition of the environment crises as a matter of international concern, requiring an integrated international response. The conference adopted a declaration on the Human Environment containing principles, which, although not binding by themselves, subsequently inspired or explicitly found their way into a large number of binding international instruments.

Two decades later, the 1992 UN Conference on Environment and Development (UNCED), held in Rio de Janeiro, constituted the second major milestone in the development of international environmental law. This conference focused on the linkage between the environment and development and the need for sustainable development, as set out in the 1987 Brundtland Report. This approach was reflected in the Rio Declaration on Environment and Development, containing 27 principles, which may be regarded as the successor to the Stockholm Declaration. One important outcome of the UNCED was the establishment in 1993 of the UN Commission on Sustainable Development. The Commission consists of 53 Member States elected for a period of three years. It reports to the UN Economic and Social Council (ECOSOC) and its mandate is to monitor the implementation of *Agenda 21* (UNCED's programme action).

These international conferences stimulated the development of international environmental law. Some of the basic principles of international environmental law are as follows:

(i) **The polluter pays principle:** According to this principle, the polluter should bear the expenses of carrying out the anti-pollution measures as decided by the public authorities. The costs of these measures would, thus, be reflected in the costs of goods and services, which cause the pollution.

(ii) **Principle of non-discrimination:** According to this principle, polluters causing transboundary pollution would be treated no less severely, than they would be, if they caused similar pollution within their own country. This right of equal access entails that a victim of transboundary pollution is granted no less favourable treatments than victims in the country where the pollution originated.

(iii) **Precautionary principle:** According to this, lack of full scientific certainty shall not be used as a reason for postponing measures to prevent environmental degradation. The actual application of this principle is the measures taken under the 'Montreal Protocol on Substances that Deplete the Ozone Layer'.

(iv) **Principle of common differentiated responsibilities:** According to this, nations should divide the costs of measures to protect the environment on the basis of the fact that they have made different contributions to global environmental degradation. One of the clearest examples of the application of this principle may be found in the Convention on Climate Change. This convention not only accepts that developing countries need to comply with less strict standards than developed countries; it also accepts that they are entitled to technological and financial assistance in order to help them meet their obligations under the treaty.

(v) **Principle of intergenerational equity:** According to this, nations are obliged to take into account the long-term effects of their actions affecting the environment. This principle attempts to emphasise that attention should not only be paid to long-distance effects but also to the long-term effects of human activity. After all, present-day decisions may restrict future uses of natural resources and may force upon future generations considerable cleaning-costs. It cannot simply be assumed that future generations will be able to develop the necessary technology for this purpose.

2.3.2 International Institutions

Now, let us touch upon some of the international institutions and agencies that were instrumental in bringing about environmental reforms through laws and legislations. These include:

- **The United Nations Environment Programme (UNEP):** Set up in the wake of the Stockholm Conference, UNEP is not a specialised agency but merely a 'programme', without executive powers and with a budget to a large extent dependent on voluntary contributions from States. UNEP's main programme consists of better coordination between existing UN programmes in the field of the environment, rather than to develop its own programmes. Nevertheless, UNEP has gradually managed to carve out a useful role for itself by taking the initiative on such issues as the control of transboundary movements of hazardous waste and the protection of the ozone layer.
- **The Food and Agricultural Organisation (FAO):** This has been active on subjects such as deforestation and conservation of fisheries.
- **The World Meteorological Organisation (WMO):** This plays a crucial role in the field of climate change and global warming.
- **The International Atomic Energy Agency (IAEA):** This agency although not initially much concerned on the subject, but in the wake of the Chernobyl disaster in Russia has focused its attention towards environmental pollution.
- **International Bank for Reconstruction and Development (IBRD)/The World Bank (WB):** This has, in recent years, begun to pay more attention to the environmental side effects of its lending policies.

Outside the UN system, regional organisations, which have adopted significant legal instruments in the field of the environment, include the Organisation for Economic Cooperation and Development (OECD), the Conference on Security and Cooperation in Europe (CSCE) and the Organisation of African Unity (OAU).

2.3.3 Key International Treaties

The following are the important international environmental treaties that deal with air, water, hazardous waste, nuclear wastes, etc.:

(i) The 1992 Helsinki Convention on the protection and use of transboundary watercourses and international lakes. (This convention deals with international watercourses.)
(ii) The 1973 Convention for the prevention of pollution from ships (MARPOL).
(iii) The 1982 UN Convention on the law of the sea (UNCLOS).
(iv) The 1992 Paris Convention for the protection of the marine environment of the North-East Atlantic. (The conventions listed at (ii) to (iv) deal with marine pollution.)
(v) The 1979 ECE Convention on long-range transboundary air pollution-protocols on sulphur dioxide (SO_2), nitrogen oxides (NO_X) and volatile organic compounds (VOCs).
(vi) The 1985 Vienna Convention for the protection of the ozone layer (Montreal Protocol and London and Copenhagen Amendments).
(vii) The 1992 UN Convention on climate change. (The conventions listed at (v) to (vii) deal with atmospheric pollution.)
(viii) The 1989 Basel Convention on the control of transboundary movements of hazardous wastes and their disposal.
(ix) The 1991 Bamako Convention on the ban of the import into Africa and the control of transboundary movement and management of hazardous wastes within Africa. (The conventions listed at (viii) and (ix) deal with hazardous waste.)
(x) The 1986 IAEA Convention on early notification of a nuclear accident.
(xi) The 1986 IAEA Convention on assistance in the case of a nuclear accident.
(xii) The 1994 IAEA Convention on nuclear safety. (The conventions listed at (x) to (xii) deal with nuclear energy.)
(xiii) The 1991 ECE Convention on environmental impact assessment in a transboundary context.
(xiv) The 1992 UN Convention on biological diversity.

2.3.4 Objectives and Principles of Legislation

The objectives of environmental legislation are to provide a set of enforceable and standard rules to contribute to the pursuit of:

(i) preserving, protecting and improving the quality of the environment;
(ii) protecting human health;
(iii) utilising natural resources in a prudent and rational way;

(iv) promoting measures at international level to deal with regional or worldwide environmental problems.

Environmental legislation seeks to regulate pollution of the natural environment in relation to air, noise, vibration, water, radiation and soil. It contains rules relating to the conservation of the natural environment, the protection of endangered species, the promotion of biological diversity, the protection of forests and the pursuit of environmentally friendly agriculture. With regard to the human environment, it seeks to protect human beings (the consumer) against contaminated food, dangerous or defective products, economic harm and danger in travel. With regard to the human-made environment it seeks the protection of historic and cultural environment. It seeks harmonisation of standards and the enforcement of those standards through legislative provisions particularly relating to the introduction of appropriate environmental management systems to harness scarce resources.

The further objectives of environmental legislation are to set standards such as achieving a high level of protection by taking into account the diversity of situations in the various regions, to give a legal basis to the precautionary principle whereby legislative action is taken where there is no reason to believe that substances or energy or materials introduced directly or indirectly into the environment, may, or are likely to create, hazards to human health, harm living resources, damage communities or interfere with other legitimate uses. This can be done even where there is no conclusive evidence of a cause or relationship between inputs and their effects and to give a legal basis to the principle that preventive action should be taken as prevention and if successful, to advise all the detailed legislation relating to steps to be taken to cure a problem.

Environmental legislation is generally based on the principles that environmental damage should, as a priority, be rectified at the source and that the polluter should pay. Environmental protection requirements should be integrated into the definition and implementation of legislation relating to non-environmental issues. In the case of cross border matters, relating to provisions primarily of a fiscal matter, measures concerning town and country planning and land use (other than waste management), management of water resources and measures significantly affecting the choice between different energy resources and the general structure of energy supply, decisions should only be adopted with unanimous agreement between the participants.

Other legal principles produced by the expert group are that countries must:

- conserve and use the environment including its natural resources for the benefit of both the present and future generations;
- maintain ecosystems and ecological processes essential for the functioning of the biosphere, and preserve biological diversity;

- observe the principle of optimum sustainable yield in use when dealing with natural resources and ecosystems;
- establish adequate environmental protection standards and monitor changes and publish relevant data on environmental quality and resource use.

LEARNING ACTIVITY 2.4

List some of the environmental legislations and international conventions and treaties related to your area of work.

Note:

a) Write your answer in the space given below.

b) Check your answer with the one given at the end of this book.

2.4 ENVIRONMENTAL LEGISLATIONS IN INDIA

India has had a long history of environmentalism with the passage and codification of acts such as the Indian Penal Code, The Criminal Procedure Code, The Bengal Smoke Nuisance Act of 1905, The Indian Motor Vehicle Act, The Factories Act, The Indian Forest Act, The Mines and Minerals (Regulation and Development) Act, The Industries (Development and Regulation) Act, The Forest (Conservation) Act and The Merchant Shipping Act.

The Indian Penal Code, passed in 1860, penalises persons responsible for causing defilement of water of a public spring or reservoir with imprisonment or fines. Traditionally, the interpretation of the Code has been viewed as a conservative attempt at enforcement. This is because punishment and fines have been characterised as meagre. In addition, fouling a 'public spring' has not, by definition, also included a 'public river', which is where most pollution occurs. Finally, the specific language of the Code places the burden of proof on the prosecution. Polluters must 'voluntarily', 'with intent', or 'knowingly' discharge damaging effluents, making successful prosecution problematic in a court of law.

The Factories Act also addresses public safety and health issues. The legislation addresses the discharge of water and effluents by factories, calling for effective arrangements for disposal at the plant level. As in the Indian Penal Code, penalties have been provided.

Section 12 of the Act empowered each State government to legislate its own rules. As a result, a number of states passed versions of the Factory Act, including Uttar Pradesh in 1950, Tamil Nadu in 1950, West Bengal in 1958, Maharashtra in 1963 and Mysore in 1969. Each tailored the Act to suit its particular situation. In Uttar Pradesh, for example, disposal of effluents had to have the approval of the State's pollution board. In Tamil Nadu, the ruling entity with similar responsibilities was the Director of Fisheries. In Maharashtra, local authorities were granted with jurisdiction in such matters.

The watershed event in the environmental movement was the Stockholm Conference on Human Environment in June 1972. The conference made it apparent to all attendees that each nation needed to adopt comprehensive legislation addressing health and safety issues for people, flora and fauna. The United Nations, organisers of the conference, requested each participant to provide a country report. Stockholm, thus, served as the genesis for the series of environmental measures India passed in the years to come.

2.4.1 Constitution of India

India was the first country to insert an amendment into its Constitution allowing the State to protect and improve the environment for safeguarding public health, forests and wildlife. The 42nd amendment was adopted in 1976 and went into effect on 3rd January 1977. The language of the Directive Principles of State Policy (Article 47) requires not only a protectionist stance by the State but also compels the State to seek the improvement of polluted environments. This allows the government to impose restrictions on potentially harmful entities such as polluting industries.

An important subtlety of the Directive's language is the provision that the article "shall not be enforceable by any court, but it shall be the duty of the State to apply these principles in making laws". This allows the Directive to be an instrument of guidance for the government, while at the same time, since no law has been passed, no individual can violate existing law.

Although State governments have clearly delineated lines of authority and jurisdiction, Article 253 of the Constitution provides the Union government with sweeping powers to implement laws for any part of India with regard to treaties made with another country or decisions made by an international body. For internal environmental matters, the Constitution provides for a distribution of legislative powers between the Union and the States. This was done by the creation of three jurisdictional listings—Union, State and Concurrent.

The central government can enact a law on any items on the Union and Concurrent lists, and, in certain cases, may enact a law on an item on the State List. To do so, the central government must approach the State Legislatures. In the case of 'water', which is an entry on the State List, Article 252 requires at least two or more State legislatures pass resolutions empowering the Parliament to pass water-related legislation. Approaching the State on issues within its jurisdiction has historically led to cumbersome delays. For example, it was almost a decade after the first draft bill was being debated, before the Parliament was able to pass India's first major water legislation in 1974. The process of State legislatures voting to empower the Parliament to pass water-related measures repeated itself during subsequent amendments to the water law in 1978 and 1988.

2.4.2 Union Government Initiatives

As early as 1962, the Union Ministry of Health had begun to address water pollution issues by appointing a study committee. The committee made recommendations for both central and state level action. Jurisdictional questions remained unsolved between the States and Union government, but by 1965 a draft bill was finally being circulated which allowed the States to pass resolutions authorising Parliament to enact legislation on their behalf. By 1969, a bill, the Prevention of Water Pollution, had been introduced. Ultimately, a modified version, i.e., the Water (Prevention and Control of Pollution) Act, was passed in 1974. Institutionalising a regulatory agency for controlling water pollution marked the first true commitment on the issue by the Indian Parliament. The Water Act also established the Pollution Control Boards at central government and state government levels.

In the aftermath of the Water Act, the then Prime Minister Indira Gandhi moved to enact a series of environmental measures. The Department of Environment (DOE), created in 1980, performed an oversight role for the Union government. DOE did environmental appraisals of development projects, monitored air and water quality, established an environmental information system, promoted research, and coordinated activities between Union, State and local governments.

Establishment of Ministry of Environment & Forest (MOEF)

The government of the then Indian Prime Minister Rajiv Gandhi recognised the deficiencies and created in 1985 the Ministry of Environment and Forests (MOEF). The MOEF was more comprehensive and institutionalised, and had a Union Minister and Minister of State, two political positions answering directly to the Prime Minister. The agency comprised eighteen divisions and two independent units, the Ganga Project Directorate and the National Mission on Wastelands Development. It continued the same functions that DOE originally had, such as monitoring and enforcement, conducting environmental assessments and surveys, but also did promotional work about the environment. In 1981, the Air (Prevention and Control of Pollution) Act was passed, and in 1986, the Parliament passed the Environment (Protection) Act, designed to act as umbrella legislation on the environment. The responsibility entrusted to administer the new legislation also fell to the Union and State pollution control boards.

In December 1993, the MOEF completed its Environmental Action Plan to integrate environmental considerations into developmental strategies, which, among other priorities, included industrial pollution reduction. In principle, though, the MOEF was already actively pursuing pollution abatement and prevention policies. Among the strategies employed by the MOEF was the implementation of a *polluter pays principle*. The MOEF imposed significant increases in its water cess, increasing it six-fold from early 1992. Since then, the MOEF has decided to increase the water cess again since the charges were still well below the current effluent treatment costs.

The MOEF also decided to shift from concentration to load-based standards. This would add to a polluter's costs and remove incentives to dilute effluents by adding water, and strengthen incentives for adoption of cleaner technologies. It also issued water consumption standards, which were an additional charge for excessive water use.

Targeting small-scale industries has been an important task as well, since these facilities greatly add to the pollution load. The Ministry provides technical assistance and limited grants to promote central effluent treatment plants. It has also created industrial zones to encourage clusters of similar industries in order to help reduce the cost of providing utilities and environmental services.

In January 1994, the Union government introduced new restrictions and prohibitions on the expansion and/or modernisation of any activity or new project, unless an environmental clearance is granted. Projects must, for example, submit an environmental impact assessment (EIA) and environmental management plan (EMP). The industry sector and the size of the project determine whether the MOEF or the State government has jurisdiction in assessing the EIA. However, in some highly polluting sectors such as power plants, cement and petrochemicals, the MOEF generally clears the EIA. In addition, State governments may issue more stringent restrictions for heavily polluted areas. One example is the Vapi region in Gujarat, where the State now has a total ban on new activities for the country's 17 most highly polluting sectors.

LEARNING ACTIVITY 2.5

List at least three government agencies responsible for environment protection in India.

Note:

a) Write your answer in the space given below.
b) Check your answer with the one given at the end of this book.

SUMMARY

In this Unit, we discussed the history of environmental management in India, learnt the impetus provided by non-government sector movements to the formulation of policies and legislations and described the institutional framework for environmental management.

REFERENCES

BSI , 1996. British Standards Institution Guide to ISO 14001: 1996, http://www.bsi.org,

Ministry of Natural Resources and Tourism (MNRT). 1998a. *National Forest Policy*. Dar es Salaam, United Republic of Tanzania. Government Printer

SUGGESTED READINGS

Carson, R. 1962. *Silent Spring*. Hamish Hamilton, London.

Meadows, D.H., Meadow, D.L., Randers, J. and Behrens, W.W., 1972. *The Limits of Growth*. Earth Island Ltd, London.

World Commission on Environment and Development, 1987. *Our Common Future*. Oxford University Press, Oxford.

http://www.envfor.nic.in Official website for Ministry of Environment and Forests, Government of India.

UNIT 3

Environmental Impact Assessment (EIA)

STRUCTURE

OVERVIEW

In Unit 1, we introduced you to some of the environmental management tools. One such tool we mentioned in that context was environmental impact assessment (EIA). In the present Unit, i.e., Unit 3, we will discuss EIA in detail. We will begin the Unit by explaining what EIA entails. We will then give you an account of EIA evolution in the world over. We will also discuss how EIA can be used to predict environmental changes and introduce strategic environmental assessment (SEA). We will close the Unit by listing some of the environmental clearance procedures in practice in India.

LEARNING OBJECTIVES

After completing this Unit, you should be able to:

- discuss environmental impact assessment (EIA) as an environmental management tool;
- trace the evolution of EIA;
- discuss what forecasting of environmental changes entails;
- explain strategic environmental assessment (SEA);
- list and comply with the environmental clearance procedures in India;
- plan and carry out an environmental impact assessment study.

3.1 INTRODUCTION TO EIA

In recent years, there has been a remarkable growth of interest in environmental issues, sustainability and the better management of development in harmony with the environment. Associated with this growth of interest has been the introduction of new legislation, emanating from national and international agencies (e.g., the European Commission) that seek to influence the relationship between development and environment. Environmental impact assessment (EIA) is an important example. It is defined as an activity designed to identify and predict the impact of legislative proposals, policies, programmes, projects and operational procedures on the bio-geophysical environment and on the health and well being of human beings and to interpret and communicate information about the impact.

That is to say, EIA focuses on problems, conflicts or natural resource constraints that could affect the viability of a project. It also examines implications of a project that might harm people, their homeland or their livelihoods, or other nearby developments. After predicting the problems, an EIA identifies measures to minimise the problems and outlines ways to improve the project's suitability for its proposed environment. In the last three decades, EIA has been recognised as the most valuable, inter-disciplinary and objective decision-making tool with respect to alternate routes for development, process technologies and project sites. It is considered an ideal anticipatory mechanism allowing measures that ensure environmental compatibility in our quest for socio-economic development.

EIA is generally wider in scope and less quantitative than other techniques, such as cost-benefit analysis. EIA has the potential to be a basis for negotiation among the developers, public interest groups and planning regulators.

3.1.1 Purposes of EIA

EIA is a process with several important purposes, which can be categorised as follows:

- **To facilitate decision-making:** For the decision-maker, for example, the local authority, it provides a systematic examination of the environmental implications of a proposed action, and sometimes alternatives, before a decision is taken. The decision-maker along with other documentation relating to the planned activity can consider the environment impact statement (EIS).

- **To aid in the formation of development:** Many developers see EIA as another set of hurdles for them to cross in order to proceed with their various activities. They may also see the process involved in obtaining the permission from various authorities as costly and time-consuming. In reality, however, EIA can be of great benefit to them, since it can provide a framework for considering location and design issues and environmental issues in parallel. It can be an aid to the formulation of developmental actions, indicating areas where the project can be modified to minimise or eliminate altogether the adverse impacts on the environment. The consideration of environmental impacts early in the planning life of a development can lead to environmentally sensitive development; to improved relations between the developer, the planning authority and the local communities; to a smoother planning permission process and sometimes to a worthwhile financial return on the expenditure incurred.
- **To be an instrument for sustainable development:** The key characteristics of sustainable development include maintaining the overall quality of life, maintaining continuing access to natural resources and avoiding lasting environmental damage. Institutional responses to sustainable development are, therefore, required at several levels. For example, issues of global concern, such as ozone-layer depletion, climate change, deforestation and biodiversity loss, require a global political commitment to action. The United Nations Conference on Environment and Development (UNCED) held in Rio de Janeiro in 1992 was an example of international concern and also of the problems of securing concerted action to deal with such issues. Governments have recognised the interaction of economic and social development and the ecosystems, and the reciprocal impact between human actions and the biogeophysical world. While there are attempts to manage this interaction better, investigations reveal disquieting trends that could have devastating consequences for the quality of the environment. These trends are likely to be more pronounced in developing countries where, because of greater rates of population growth and lower current living standards, there is more pressure on environmental resources.

In short, an interaction among the resources, sectors and policies is necessary for sustainable development as illustrated in Fig. 3.1 and EIA contributes to this process.

3.1.2 Steps in EIA Process

EIA represents a systematic process that examines the environmental consequences of the development actions, in advance. The emphasis of an EIA is on prevention

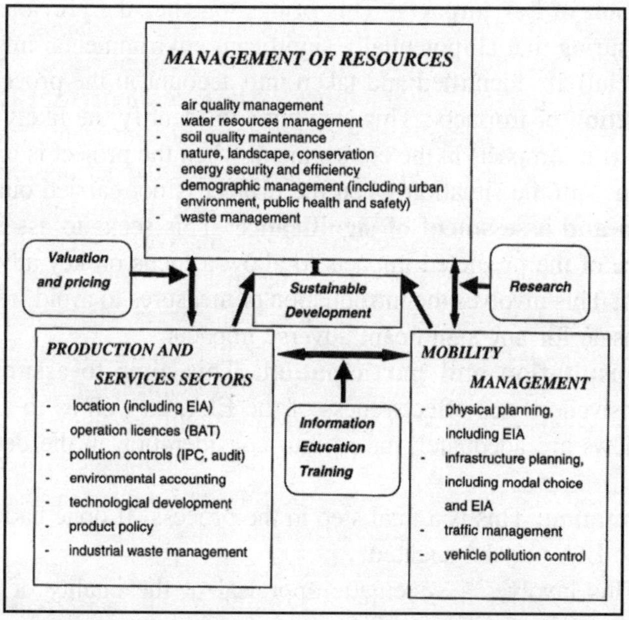

Figure 3.1 Sustainable Development: An Illustration

and, therefore, is more proactive than reactive in nature. The EIA process involves a number of steps, some of which are listed below:

- **Project screening:** This entails the application of EIA to those projects that may have significant environmental impacts. It is quite likely, however, that screening is done partly by the EIA regulations, operating in a country at the time of assessment.
- **Scoping:** This step seeks to identify, at an early stage, the key, significant environmental issues from among a host of possible impacts of a project and all the available alternatives.
- **Consideration of alternatives:** This seeks to ensure that the proponent has considered other feasible approaches, including alternative project locations, scales, processes, layouts, operating condition and the *no-action* option.
- **Description of the project/development action:** This step seeks to clarify the purpose and rationale of the project and understand its various characteristics, including the stages of development, location and processes.
- **Description of the environmental baseline:** This includes the establishment of both the present and future state of the environment, in the absence of the project, taking into account the changes resulting from natural events and from other human activities.

- **Identification of key impacts:** This brings together the previous steps with a view to ensuring that all potentially significant environmental impacts (adverse and beneficial) are identified and taken into account in the process.
- **The prediction of impacts:** This step aims to identify the likely magnitude of the change (i.e., impact) in the environment when the project is implemented in comparison with the situation when the project is not carried out.
- **Evaluation and assessment of significance:** This seeks to assess the relative significance of the predicted impacts to allow a focus on key adverse impacts.
- **Mitigation:** This involves the introduction of measures to avoid, reduce, remedy or compensate for any significant adverse impacts.
- **Public consultation and participation:** This aims to assure the quality, comprehensiveness and effectiveness of the EIA, as well as to ensure that the public's views are adequately taken into consideration in the decision-making process.
- **EIS presentation:** This is a vital step in the process. If done badly, much good work in the EIA may be negated.
- **Review:** This involves a systematic appraisal of the quality of the EIS, as a contribution to the decision-making process.
- **Decision-making:** At this stage, decisions are made by the relevant authority of the EIS (including consultation responses) together with other material considerations as to whether to accept, defer or reject the project.
- **Post-decision monitoring:** This involves the recording of outcomes associated with development impacts, after the decision to proceed with the project. It can contribute to effective project management.
- **Auditing:** This follows monitoring and involves comparing actual outcomes with predicted outcomes, and can be used to assess the quality of predictions and the effectiveness of mitigation. It provides a vital step in the EIA learning process.

Figure 3.2 illustrates the steps involved in the EIA process.

Note that the actual EIA process is not so linear and sequential as Fig. 3.2 seems to suggest. In other words, it is a cyclical process involving feedback and interaction among the various steps and the sequence of the steps may also vary. Note also that we will discuss the steps involved in detail in Unit 4.

3.1.3 Hierarchy in EIA

The EIA studies are broadly categorised as:

(i) **Site selection studies:** These studies involve an evaluation of the alternative sites with respect to environmental and project attributes such as proximity to raw materials, infrastructure facilities, markets, etc. These studies aim at ranking site alternatives for objective decision-making.

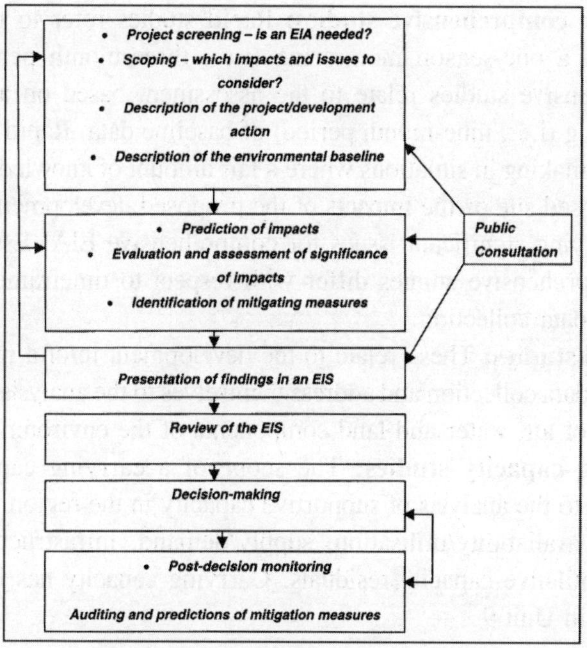

Figure 3.2 Steps in an EIA Process

LEARNING ACTIVITY 3.1

Define the EIA process.

Note:

a) Write your answer in the space given below.

b) Check your answer with the one given at the end of this book.

(ii) **Rapid or comprehensive studies:** Rapid studies refer to the assessment based on a one-season monitoring (i.e., three-month period), whereas comprehensive studies relate to the assessment based on a three-seasons monitoring (i.e., nine-month period) of baseline data. Rapid EIA facilitates decision-making in situations where a fair amount of knowledge exists about the proposed site or the impacts of the proposed development. It also helps in identifying significant issues for comprehensive EIA. Essentially, rapid and comprehensive studies differ with respect to timeframes required for baseline data collection.

(iii) **Regional studies:** These relate to the development in/of a region based on seasonal data collection and address themselves to the analysis of assimilative capacity of air, water and land components of the environment.

(iv) **Carrying capacity studies:** The scope of a carrying capacity study is extended to the analysis of supportive capacity in the region with respect to resource availability/utilisation, supply/demand, infrastructure/congestion and assimilative capacity/residuals. Carrying capacity has been discussed in detail in Unit 9.

In the last two decades, national governments and also financial institutions have realised that EIA has to be an integral part of the project life cycle: from project conceptualisation to post implementation corrective action. Figure 3.3 illustrates this cycle.

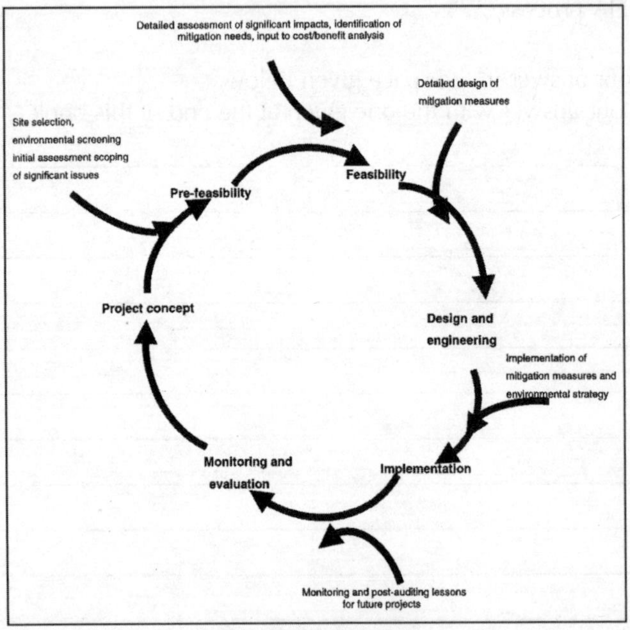

Figure 3.3 EIA Cycle

An EIA exercise culminates in an environmental impact statement (EIS), which we will study, next.

3.1.4 Environmental Impact Statement (EIS)

The environmental impact statement (EIS) provides documentation of the information and estimates derived from the various steps in the EIA process. The information contained in an EIS provides the decision-makers/regulators with valuable information that could ultimately contribute to either the abandonment or substantial modification of a proposed development action. A typical EIS contains the following three parts:

- **Part 1 – Methods and key issues:** This part deals with the statement of methods used and a summary of key issues.
- **Part 2 – Background to the proposed development:** This part deals with preliminary studies (i.e., need, planning, alternatives, site selection, etc.), site description/baseline conditions, description of proposed development and construction activities and programmes.
- **Part 3 – Environmental impact assessments on topic areas:** This part deals with land use, landscape and visual quality, geology, topography and soils, hydrology and water quality, air quality and climate, terrestrial and aquatic ecology, noise, transport, socio-economic and interrelationships between effects.

LEARNING ACTIVITY 3.2

Write three ways in which EIA will help in Business Management.

Note:

a) Write your answer in the space given below.

b) Check your answer with the one given at the end of this book.

3.1.5 Impact Indicators

An impact indicator is an element or a parameter that provides a measure (in at least some qualitative sense) of the significance of the effect, i.e., the magnitude of an environmental impact. Some indicators such as morbidity and mortality statistics and crop yields have associated numerical scales. Other impact indicators, however, can only be ranked as 'good', 'better', 'best' or 'acceptable', 'unacceptable', etc. The selection of a set of indicators is often a crucial step in the impact assessment process, requiring input from the decision-maker. In the absence of relevant goals or policies, the assessor himself or herself may suggest some indicators and scales, but he or she should not proceed with the assessment until his or her proposals are accepted.

The most widely used impact indicators are those within statutory laws, acts, i.e., indicators such as air and water quality standards that have statutory authority. For example, the problem of designing an environmentally acceptable oil-fired generating station is simplified for the engineers, if they are given one or both of the following:

- Emission standards for various pollutants.
- Air and water quality standards.

These standards integrate the worth that a jurisdiction places on clean air and clear water. The numerical values that have been derived from examination of the available toxicological matter are data relating polluting dosages to health and vegetation effects, combined with a consideration of the best practical technology. Factors such as the displacement of arable land by industry are also equally important. An EIA that ignores these other components is incomplete and sometimes misleading (Munn, 1979).

3.2 EVOLUTION OF EIA

To understand the use of EIA as a tool for environmental management, let us discuss how EIA has evolved over the years.

3.2.1 Evolution of EIA Worldwide

United States of America was the first country to assign mandatory status to EIA through its National Environmental Protection Act (NEPA) of 1969. A host of industrialised countries have since implemented EIA procedures. Canada, Australia, the Netherlands and Japan adopted EIA legislation in 1973, 1974, 1981 and 1984, respectively. In July 1985, the European Community (EC) issued a directive making environmental assessments mandatory for certain categories of projects (Wood, 1994).

Among the developing countries, Columbia was the first Latin American country to institute a system of EIA in 1974. In Asia and the Pacific region, Thailand and the Philippines have long established procedures for EIA. EIA was made mandatory in Sri Lanka in 1984. The EIA process in Africa is sketchy, although a number of nations including Rwanda, Botswana and Sudan have some experience of EIA (Wathern, 1988).

Bilateral and multilateral agencies have also recognised the value of EIA as a decision-making tool. The Organisation for Economic Cooperation and Development (OECD) issued recommendations on EIA to its constituent States in 1974 and 1979, and for development aid projects in 1986. OECD issued guidelines for good practices in EIA in 1992 (OECD, 1992). United Nations Environment Programme (UNEP) in 1980 provided guidance on EIA of the development proposals (UNEP, 1980) and supported research on EIA in developing countries (Ahmad and Swamy, 1985). UNEP, in 1987, set out goals and principles of EIA for the member countries and provided guidance on basic procedures for EIA in 1988.

The World Conservation Strategy pinpointed the need to integrate environmental considerations with development in 1980 (IUCN, 1980). EIA became an integral part of World Bank policy in 1987 which states that environmental issues must be addressed as part of overall economic policy. In 1989, the World Bank issued the Operational Directive on Environmental Assessment (O.D. 4.00), which was revised and updated in October 1991 (O.D. 4.01). Asian Development Bank in 1990 published guidelines for EIA (ADB, 1990). Importance of EIA was echoed in the Brundtland Report (WCED, 1987), and at United Nations Earth Summit on environment and development held at Rio de Janeiro in 1992 (UNCED, 1992). As foreseen by Garner and O'Riodan (1982), development of EIA, as a tool for decision-making world-over, has emerged through the following stages:

- No formal accounting, decisions made on interest group lobbying and engineering feasibility; primary emphasis on economic development.
- Conventional cost-benefit analysis; emphasis on efficiency criterion and engineering feasibility; major concern still on economic development.
- Innovative cost-benefit analysis, use of multiple objectives and discount rates, imaginative proxy pricing mechanisms; economic development as one of the objectives.
- EIA mainly concerned with describing the repercussions of the proposals on biophysical processes; economic development still primary objective.
- EIA with more attention paid to socio-cultural as well as biophysical systems, economic development but not the sole objective.

The summary of evolution of EIA in various countries is presented in Table 3.1.

Table 3.1 Evolution of EIA Worldwide

Australia	Environmental Protection (Impact of Proposals) Act 1974, Commonwealth of Australia
Bangladesh	No specific EIA legislation, however there was a Declaration that Environmental Impact Assessments should be carried out for all major development projects, 1995
China	Environmental Protection Law, 1979
USA (California)	California Environmental Quality Act (CEQA) of 1971
Canada	Federal Environmental Assessment and Review Process Guidelines Order 1984, Canada
France	Law on Protection de la Nature, 1978
India	Notifications dated May 5, 1994 under the Environment Protection Act, 1986
Japan	Principles for Implementing EIA by Environmental Agency, 1984
Malaysia	Environmental Quality (Prescribed Activity) (EIA) Order, 1987
New Zealand	Resource Management Act 1991, New Zealand
The Philippines	Presidential Decree (PD) 1151 Philippines Environment Policy, 1975
	PD 1586 Establishing the Environmental Impact Statement (EIS), 1978 Rules and Regulations to Implement the EIS System, 1987
Sri Lanka	National Environmental Act 1980, amended in 1986
Thailand	Improvement and Conservation of National Environmental Quality Act 1975, amended in 1978
The Netherlands	EIA Policy, 1986
United States	US Environmental Policy Act, 1969
Vietnam	Environmental Protection Law, 1994
Western Australia	Environmental Protection Act 1986
West Germany	Cabinet Resolution, 1975

Let us now work out Learning Activity 3.3.

LEARNING ACTIVITY 3.3

List three Asian countries where law requires EIA and also name the relevant law.

Note:

a) Write your answer in the space given below.

b) Check your answer with the one given at the end of this book.

3.2.2 Evolution of EIA in India

EIA in India was started in 1976-77, when the Planning Commission asked the then Department of Science and Technology to examine the river-valley projects from the environmental angle. This was subsequently extended to cover those projects, which required approval of the Public Investment Board. These were administrative decisions, and lacked the legislative support. The Government of India enacted the Environment (Protection) Act on 23rd May 1986. To achieve the objectives of the Act, one of the decisions taken was to make EIA statutory. After following the legal procedure, a notification was issued on 27th January 1994 and subsequently amended on 4th May 1994, 10th April 1997 and 27th January 2000 making environmental impact assessment statutory for 30 activities. This is the principal piece of legislation governing EIA in India. Besides this, the Government of India under Environment (Protection) Act 1986 issued a number of notifications, which are related to environmental impact assessment. These are limited to specific geographical areas, and are summarised below:

- Prohibiting location of industries except those related to Tourism in a belt of 1 km from high tide mark from the Revdanda Creek up to Devgarh Point (near Shrivardhan) as well as in 1 km belt along the banks of Rajpuri Creek in Murud Janjira area in the Raigarh district of Maharashtra (6th January 1989).
- Restricting location of industries, mining operations and regulating other activities in Doon Valley (1st February 1989).
- Regulating activities in the coastal stretches of the country by classifying them as coastal regulation zone and prohibiting certain activities (19th February 1991).
- Restricting location of industries and regulating other activities in Dahanu Taluka in Maharashtra (6th June 91).
- Restricting certain activities in specified areas of Aravalli Range in the Gurgaon district of Haryana and Alwar district of Rajasthan (7th May 1992).
- Restricting industrial and other activities, which could lead to pollution and congestion in the north west of Numaligarh in Assam (July 1996).

LEARNING ACTIVITY 3.4

List three major developmental projects in India funded by the World Bank, which required EIA and the main aspect assessed in each of these projects.

Note:

a) Write your answer in the space given below.

b) Check your answer with the one given at the end of this book.

3.3 FORECASTING ENVIRONMENTAL CHANGES

An EIA should be able to, among others, predict the nature and extent of the impact of human activities on the environment. Table 3.2 gives a list of human-induced environmental changes, which can be either benign or malignant to the environment.

Of importance here are not only estimates of changes in environmental quality but also estimates of rate of change. A slow change may be acceptable, especially if it leads to a new stability, whereas rapid change or large fluctuations may place intolerable burdens on ecosystems. Of equal or perhaps greater importance is the degree of irreversibility of an environmental change, which will be either absolute, as in the extinction of a species, or partly absolute in that the situation can only be

Table 3.2 Environmental Changes

Medium	*Changes and Rates of Change in*
Soil	Quality (e.g., depth, structure, fertility, degree of salinisation or acidification, etc.)
	Stability
	Area of arable land
Air	Quality
	The climatic elements
Water	Quantity
	Quality
	Seasonality
	Area of human-made lakes
	Extent of irrigation canals
Biota	Abundance/scarcity of species or genetic resources
	Extent of crops, ecosystems, vegetation and forests
	Diversity of species
	Extent of provision of nesting grounds, etc., for migratory species
	Abundance/scarcity of pests and disease organisms.

reversed over long periods of time or with unacceptable expenditures of money and energy, as in the case of catastrophic erosion.

A typical EIA contains information on the following three areas, as they relate to environmental effects:

(i) A determination of the initial reference state.
(ii) An estimate of the future state without action.
(iii) An estimate of the future state with action.

We will describe each of these, next.

Establishment of the Initial Reference State

An assessment of environmental change pre-supposes knowledge about the present state. It will be necessary, therefore, to select attributes that may be used to estimate this state. Some of these will be directly measurable; others will only be capable of being recorded within a series of defined categories, or ranked in ascending or descending order of approximate magnitude. Difficult decisions need to be made about the *population* (i.e., in a statistical sense), which is to be represented by the measured variables, and the extent to which the subdivision of this population into geographical regions, ecosystems, etc., is either feasible or necessary. In fact, it must be emphasised that the establishment of an initial reference state is difficult because not only are environmental systems dynamic but also they contain cyclical and random components.

Predicting the Future State in the Absence of Action

In order to provide a fair basis for examining the impact of human activities on the environment, an EIA must estimate the future environmental states in the absence of action. As an example, the population of a species of animal or fish may already be declining, due to over-grazing or over-fishing, even before a smelter is built. This part of analysis is largely a scientific problem, requiring skills drawn from many disciplines. The prediction will often be uncertain but the degree of uncertainty should be indicated at least in qualitative terms. For example, forecasting of droughts two or three years in advance is not yet possible, although the statistical probability that a drought (of a given severity) will occur sometime in the next hundred years can be estimated with some confidence. The decision-maker should be aware of the degree of uncertainty, which surrounds the predicted state of the environment, and have some understanding of the methods by which this uncertainty is calculated.

Predicting the Future State in the Presence of Action

For each of the proposed actions, and for admissible combinations of these actions, there will be an expected state of the environment, which is to be compared with the expected state in the absence of action. Consequently, predictions similar to those outlined above must be derived for each of the proposed alternatives.

LEARNING ACTIVITY 3.5

List at least three adverse impacts associated with highway, hydroelectric and thermal power projects.

Note:

a) Write your answer in the space given below.

b) Check your answer with the one given at the end of this book.

Table 3.3 presents the main areas of concern that may affect human beings with regard to forecasting the environmental state in the presence of actions.

Table 3.3 Areas of Human Concern

Areas of human concern	*Impact Categories*
Economic and occupational status	Displacement of population; relocation of population in response to employment opportunities; services and distribution patterns; property values.
Social pattern or life style	Resettlement; rural depopulation; change in population density; food; housing; material; agricultural; rural; urban.
Social amenities and relationships	Family life styles; schools; transportation; community feelings; participation vs. alienation; recreation; language.
Psychological features	Involvement; expectations; stress; frustration; commitment.
Physical amenities (intellectual, cultural, aesthetic and sensual)	National parks; wildlife; art galleries; archaeological monuments; wilderness; clean air and water.
Health	Changes in health; medical services; medical standards.
Personal security	Freedom from molestation; freedom from natural disasters.
Regional and traditional beliefs	Symbols; taboos; values.
Technology	Security; hazards; safety measures; benefits; emission of wastes; congestion; density.
Cultural	Leisure; new values; heritage; traditional and religious rites.
Political	Authority; level and degree of involvement; priorities; structure of decision-making; responsibility and responsiveness; resource allocation; local and minority interests; defence needs.
Legal	Restructuring of administrative management; changes in taxes; public policy.
Aesthetic	Visual physical changes; moral conduct; sentimental values.
Statutory laws and acts	Air and water quality standards; safety standards; national building acts; noise abatement by-laws.

Note that the nature of impact listed in Table 3.3 is likely to vary from place to place and from time to time, and there will be overlaps between classes (e.g., health depends partly on economic and occupational status).

In this Section, we explained the anthropogenic effect and their impacts in terms of environmental preservation. However, a specific parameter is necessary to provide a measure of the significance of an effect.

LEARNING ACTIVITY 3.6

Identify three stages in the implementation of a highway project and the environmental concerns at those stages.

Note:

a) Write your answer in the space given below.

b) Check your answer with the one given at the end of this book.

So far, we dealt with EIA, which is an indispensable tool for environmental engineers and managers alike. Now, let us introduce you to a new concept called *strategic environmental assessment*, which covers policies, plans and programmes at critical stages of development.

3.4 STRATEGIC ENVIRONMENTAL ASSESSMENT (SEA)

One of the most recent trends in EIA is its application at earlier, more strategic stages of development at the level of policies, plans and programmes, and is known as strategic environmental assessment (SEA). SEA is defined as the formalised, systematic and comprehensive process of evaluating the environmental impacts of a policy, plan or programme (PPP) and its alternatives, including the preparation of a written report on the findings of that evaluation, and using the findings in publicly accountable decision-making (Therivel et al., 1992). In other words, the EIA of policies, plans and programmes, keeping in mind that the process of evaluating environmental impacts at a strategic level, is not necessarily the same as that at a project level. In theory, PPPs are tiered—a policy provides a framework for the establishment of plans, plans provide frameworks for programmes and programmes lead to projects. The EIAs for these different PPP tiers can themselves be tiered as shown in Fig. 3.4, and so the issues at higher tiers need not be reconsidered as the lower tiers.

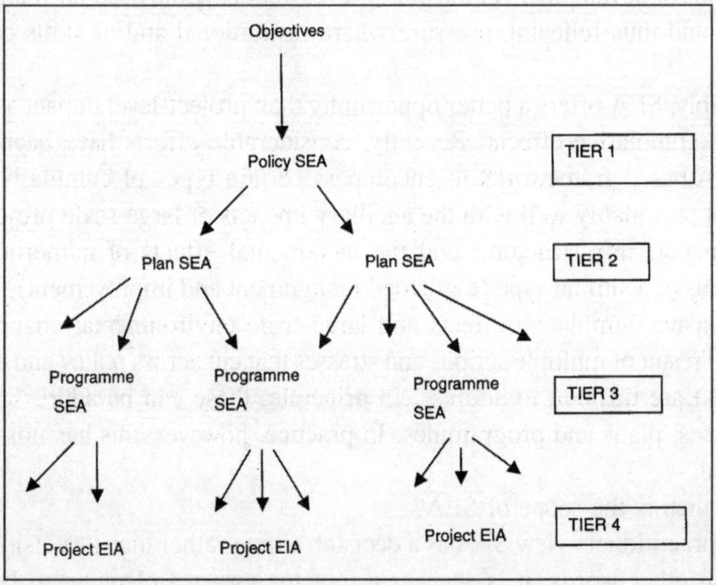

Figure 3.4 Tiers in SEAs

3.4.1 Rationale and Scope

In broad terms, the rationale for SEA of policies, plans and programmes falls into three main categories: strengthening project EIA; advancing the sustainability agenda; and addressing cumulative and large-scale effects (Jacobs and Sadler, 1989; Lee and Walsh, 1992; Sadler, 1994; Sadler and Verheem, 1996) (*http://www.nssd.net/pdf/IIEDO2.pdf*). Let us discuss these, next.

The EIA practice is constrained by certain limitations and weaknesses. These include structural weaknesses centred on the relatively late stage at which EIA is usually applied in decision-making. Put differently, high-order questions of whether, where and what type of development should take place are decided, often with little or no environmental analysis. Project-by-project EIA is also an ineffective means of examining these issues. SEA, or an equivalent approach, can be used as a complement to project-level EIA to incorporate environmental considerations and alternatives directly into policy, plan and programme design. Thus, when applied systematically in the upstream part of the decision cycle and to the economic, fiscal and trade policies that guide the overall course of development, SEA can be a vector for a sustainability approach to planning and decision-making (Brundtland Commission of WCED, 1987 and *Agenda 21* of UNCED, 1992). This upstream approach can also help in making EIA projects more consequential and reducing the time and effort involved in their preparation. SEA may yield other significant benefits, as well. For example, by ruling out certain kinds of development at the policy level, reducing the need for many project-level EIA and thus relieving pressure where institutional and/or skills capacity is limited.

Arguably, SEA offers a better opportunity than project-level impact assessment to address cumulative effects. Recently, considerable efforts have been made to extend EIA-based frameworks to encompass certain types of cumulative effects. These deal reasonably well with the ancillary impacts of large-scale projects (e.g., dams, transport infrastructure) and the incremental effects of numerous, small-scale actions of a similar type (e.g., road realignment and improvement). However, more pervasive cumulative effects and large-scale environmental change (which are the end result of multiple actions and stresses that cut across policy and ecological boundaries) are difficult to address. In principle, these can be addressed best by SEA policies, plans and programmes. In practice, however, this has not proved to be the case.

What then is the scope of SEA?

Most practitioners view SEA as a decision aiding rather than a decision-making process. In other words, it is seen as a tool for forward planning to be flexibly applied at various stages of the policy-making cycle. Under this broad perspective, SEA encompasses assessments of both broad policy initiatives and more concrete programmes and plans that have physical and spatial references (e.g., town and regional plans, regional development programmes, etc.). With this scope of coverage, one problem becomes evident, and that is, the methodologies to be applied at the opposite ends of the decision-making spectrum differ markedly. However, the principles of EIA apply at all levels.

Table 3.4 compares EIA and the evolving processes of SEA.

SEA process is as comprehensive as EIA, if not as exhaustive. We will discuss this, next.

3.4.2 SEA Process

In project EIA, impact mitigation, i.e., avoiding or reducing the project's impacts, restoring the affected environment or compensating for adverse effects, is often considered a separate stage in the process. In SEA, instead, the focus of the project is on reconsidering the PPP from a cross-cutting perspective, leading to an improved understanding of the PPP and possibly changes to the PPP: each

Table 3.4 Comparison of EIA and SEA

EIA	*SEA*
Is usually reactive to a development proposal.	Is proactive and informs development proposals.
Assesses the effect of a proposed development on the environment.	Assesses the effect of a policy, plan or programme on the environment, or the effect of the environment on development needs and opportunities.
Addresses a specific project.	Addresses areas, regions or sectors of development.
Assesses direct impacts and benefits.	Assesses cumulative impacts and identifies implications and issues for sustainable development.
Focuses on the mitigation of impacts.	Focuses on maintaining a chosen level of environmental quality.
Narrow perspective and a high level of detail.	Wide perspective and a low level of detail to provide a vision and overall framework.
Focuses on project-specific impacts.	Creates a framework against which impacts and benefits can be measured.

stage considers whether and how the PPP can be changed and improved. These changes mostly involve rewriting the PPP to minimise any negative environmental/ sustainability impacts but could also involve establishing management guidelines for the implementation of the PPP, placing constraints on lower-tier PPP. (For instance, establishing criteria for identifying future developments away from sensitive sites or requiring SEA/EIA for lower-tier PPPs and projects). Or, developing environmentally beneficial shadow PPPs or projects. As such, mitigation in SEA is an ongoing process as illustrated in Fig. 3.5:

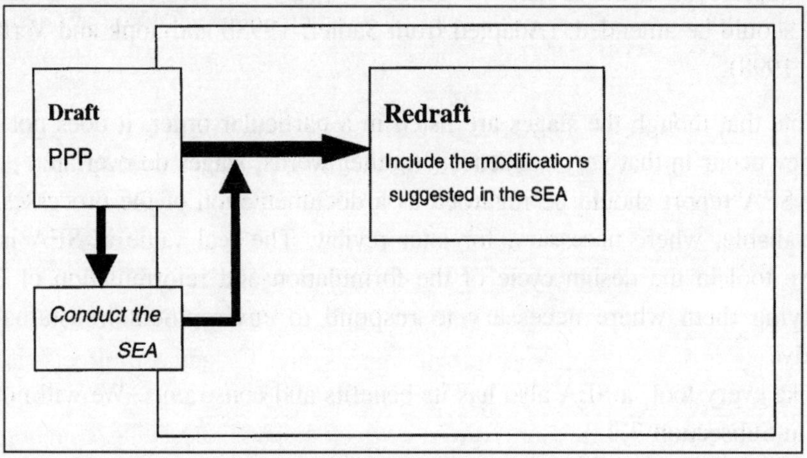

Figure 3.5 Mitigation in SEA

A SEA process involves the following stages:

(i) **Screening:** At this stage, responsible agencies carry out an appropriate assessment of all strategic decisions with significant environmental consequences.

(ii) **Timing:** At this stage, results of the assessment are available sufficiently early for use in the preparation of the strategic decision.

(iii) **Environmental scoping:** At this stage, all relevant information is provided to judge whether an initiative should proceed and objectives could be achieved in a more environmentally friendly way (i.e., through alternative initiatives or approaches).

(iv) **Other factors:** At this stage, sufficient information is available on other factors, including socio-economic considerations, either parallel to, or integrated in, the assessment.

(v) **Review:** At this stage, the quality of the process and information is safeguarded by an effective review mechanism.

(vi) **Participation:** At this stage, sufficient information on the views of all legitimate stakeholders (including the public affected) is available early enough to be used effectively in the preparation of the strategic decision.

(vii) **Documentation:** At this stage, results are identifiable, understandable and available to all parties affected by the decision.

(viii) **Decision-making and accountability:** At this stage, it should be clear to all stakeholders and all parties affected how the results were taken into account in decision-making.

(ix) **Post-decision:** At this stage, sufficient information on the actual impacts of implementing the decision is gained to judge whether or not the decision should be amended. (Adapted from Sadler, 1998b and Tonk and Verheem, 1998).

Note that though the stages are listed in a particular order, it does not mean that they occur in that very sequence. In other words, stages do overlap.

A SEA report should be regarded as a documentation of the processes used and available, where necessary, for later review. The real value in SEA is as a creative tool in the design cycle of the formulation and reformulation of PPPs, modifying them where necessary to respond to environmental/sustainability objective.

Like every tool, a SEA also has its benefits and constraints. We will now list them in Subsection 3.4.3.

LEARNING ACTIVITY 3.7

Name five stakeholders in a highway project.

Note:

a) Write your answer in the space given below.

b) Check your answer with the one given at the end of this book.

3.4.3 Benefits and Constraints of SEA

Some of the benefits of SEA include the following:

(i) promoting integrated environment and development decision-making;

(ii) facilitating the design of environmentally-sustainable policies and plans;

(iii) providing for consideration of a larger range of alternatives than is normally possible in project EIA;

(iv) taking account, where possible, of cumulative effects (particularly by focussing on the consequences of sectoral or regional-level developments) and global change;

(v) enhancing institutional efficiency (particularly, where EIA related skills, operational funds and institutional capacities are limited) by obviating the need for unnecessary project-level EAs;

(vi) strengthening and streamlining project EA by incorporating environmental goals and principles into policies, plans and programmes that shape individual projects; identifying in advance the impacts and information requirements; resolving strategic issues and information requirements and reducing time and effort taken to conduct reviews;

(vii) providing a mechanism for public engagement in discussions relevant to sustainability at a strategic level.

Some of the constraints of SEA include the following:

- A level of institutional maturity is necessary, which allows for effective inter-sectoral dialogue, for environmental considerations to be taken into account in formulating, revising and implementing policies, plans and programmes effectively, and to influence decision-making.
- Appropriate skills are needed, within government departments/agencies and private sectors (e.g., industry, environmental consulting companies) and amongst academics and NGOs.
- There is a need for adequate capacity in these sectors (both human and financial resources).

In practice, the extent to which the benefits of SEA are achieved will also depend on a number of other important factors such as:

- provisions made for SEA, e.g., legal versus administrative;
- prior record of implementation and acceptance by decision-makers;
- degree to which overall strategies of sustainable development are in place;
- scope and level(s) of process application; with the broadest range of benefits being gained from SEA systems that include review of policies as well as plans and programmes. (Adapted from Dalal-Clayton and Sadler, 1995 and Sadler and Baxter, 1997.)

3.5 ENVIRONMENTAL CLEARANCE PROCEDURE IN INDIA

As the utility of EIA became clear, there was need to establish project clearance procedure. The first step in that direction was to define the EIA process. The EIA process in India is made up of the following phases:

- Screening.
- Scoping and consideration of alternatives.
- Baseline data collection.
- Impact prediction.
- Assessment of alternatives, delineation of mitigation measures and environmental impact statement.
- Public hearing.
- Environment management plan (EMP).
- Decision-making.
- Monitoring the clearance conditions.

The Ministry of Environment and Forests (MOEF) has published guidelines for different sectors, which outlines the significant issues to be addressed in the EIA studies. In general, the following impacts of the project need to be assessed:

- **Air:** The changes in ambient levels and ground level concentrations due to total emissions from point, line and area sources, effects on soils, materials, vegetation and human health are to be assessed.
- **Noise:** The changes in ambient levels, due to noise generated from equipment and movement of vehicles, and their impact on fauna and human health are to be assessed.
- **Water:** The availability to competing users, changes in quality, sediment transport and ingress of saline water are to be assessed.
- **Land:** The changes in land use and drainage pattern, land quality including effects of waste disposal, shoreline/riverbank and their stability are to be assessed.
- **Biological:** The level of deforestation/tree-cutting and shrinkage of animal habitat, the impact on fauna and flora (including aquatic species, if any) due to contaminants/pollutants and the impact on rare and endangered species, endemic species, and migratory path/route of animals are to be assessed, as also the impact on breeding and nesting grounds.
- **Socio-economic:** The impact on the local community including demographic changes, economic status, human health and increased traffic are to be assessed.

For every project, possible alternatives need to be identified and environmental attributes compared. The alternatives identified must cover project location and process technologies including the no-project option and the alternatives need to be ranked for selection on the basis of optimum economic benefits to the community at large.

Once the alternatives have been reviewed, a mitigation plan supplemented with an environmental management plan (EMP) needs to be drawn up for the selected option to guide the proponent towards environmental improvements. Note that the EMP is a crucial input to monitoring the clearance conditions, and therefore, it must contain the details of monitoring. An EIA report, thus, needs to provide clear information to the decision-maker on the different environmental scenarios without the project, with the project and with project alternatives. This includes uncertainties. MOEF has issued sectoral guidelines and environmental appraisal questionnaires for obtaining the clearance and seeks the following documents:

- Filled-in application form (as per Schedule II of EIA Notification).
- A summary of the project/feasibility report (one copy).
- EIA (EIS)/EMP report (20 copies).
- Risk analysis on on-site emergency preparedness plan (20 copies) in case of projects involving hazardous substances.
- Site clearance from MOEF for site-specific projects mentioned in the EIA notification.

- Consent to establish from SPCB.
- NOC from the local authorities (e.g., District Collector).
- Commitment regarding the availability of water and electricity from the appropriate agencies.
- Approval of the Chief Controller of Explosives under the Petroleum Act and Rules for layout and storage of hazardous substances and from the Directorate of Industrial Safety and Health under the Factories Act and Rules.
- Comments/Observations/Recommendations of the Chief Wildlife Warden in case a wildlife habitat/migration path exists within 25 km of project site.
- Comprehensive summary rehabilitation plan, where displacement of more than 1000 people is anticipated.
- Copy of the application forwarded to the state government, in case of diversion of forestland.
- Copy of the application forwarded to the state government in case the CRZ notification applies.
- Clearance from the Airport Authority of India, if applicable.
- Details of the public hearing conducted by SPCB and copies of the advertisements issued for public hearing.
- Filled-in environmental appraisal questionnaires issued by MOEF, along with the attachments (mentioned in the questionnaire).

MOEF has issued different questionnaires for different projects, and the law requires that the public must be informed and consulted on a proposed development after the completion of EIA report. Any one is entitled to have access to the executive summary of the EIA, and the affected persons such as bonafide local residents, local associations, environmental groups active in the area and any other person located at the project site/sites of displacement must be given an opportunity to make oral/written suggestions to the State Pollution Control Board (SPCB).

The decision-making process involves consultation between the project proponent (assisted by a consultant) and the impact assessment authority (assisted by an expert group, if necessary). The decision on environmental clearance is arrived at through a number of steps including evaluation of EIA and the environmental management plan (EMP). In India, the project proponent during the project planning stage decides the type of projects, i.e., new establishment, expansion or modernisation. Later, the project proponent prepares a detailed project report/feasibility report and submits to the authorities concerned the executive summary containing the project details and findings of the EIA study, which is to be made available to the concerned public.

Where a proponent is required to obtain environmental clearance, the Impact Assessment Agency (IAA) evaluates and assesses the EIA report. In this process, the project proponent is given a chance to present his or her proposal. If a project is accepted, the IAA prepares a set of recommendations and conditions for its implementation based on this assessment. Environmental clearance conditions and recommendations of IAA are made available to the public on request through SPCB and through a web site <*http://envfor.nic.in*>. During the implementation and operation of the project, the IAA is also responsible for the environmental monitoring process.

LEARNING ACTIVITY 3.8

List three stakeholders who should be consulted during the public hearing of a coal-based thermal power plant project.

Note:

a) Write your answer in the space given below.

b) Check your answer with the one given at the end of this book.

In short, an environmental impact assessment is an integral part of any project planning and must be understood with respect to the requirement of each project. A project proponent has to first determine whether the activity requires environmental clearance and undertake EIA studies to meet the requirements of MOEF. In case of joint venture projects (with a foreign company) and/or for seeking funds from an international financial institution, an environment impact statement has to conform to the environmental assessment guidelines. You must have a fairly good knowledge of, and must be capable of anticipating, the requirements and implications on project schedules and cost. It is an inescapable fact today that an EIA report needs to be prepared in a manner that conforms to national and international regulations and guidelines. It must be seen as a proactive, rather than a purely conformist document.

SUMMARY

In this Unit, we introduced you to the concept of environmental impact assessment (EIA). We began the Unit by giving an overview of what EIA entails, and in this context, touched upon the purpose of EIA, the processes involved in EIA, etc. Subsequently, we gave an account of the evolution of EIA both in India and in other countries. We also discussed how prediction or forecasting of environmental change is vital for EIA. We then discussed strategic environmental assessment (SEA) and described the overlaps and differences between SEA and EIA. Finally, we closed the Unit by describing some of the environmental clearance procedures in practice in India.

REFERENCES

ADB, 1990. Guidelines for Environmental Impact Assessment

Dalal-Clayton D.B. and Sadler B., 1995. *Strategic Environmental Assessment. A Briefing Paper*. IIED, London.

Garner J. F. and O'Riordan T.,1982. Environmental Impact Assessment in the context of Economic Recession, The Geographical Journal, Vol. 148, No. 3, pp. 343-355, Blackwell Publishing on behalf of the Royal Geographical Society (with the Institute of British Geographers)

Jacobs P. and Sadler B. (eds), 1989. *Sustainable Development and Environmental Assessment: Perspectives on Planning for a Common Future*, Canadian Environmental Assessment Research Council, Ottawa.

Lee, N. and Walsh, F., 1992. Strategic Environmental Assessment: An Overview, *Project Appraisal*, V7, N. 3, 126-136.

Munn R.E., 1979. Environmental Impact Assessment: Principles and Procedures. The Scientific Committee on Problems of the Environment (SCOPE), Report no 5, Second Edition, Pp 190, John Wiley and Sons Chichester, England.

NEPA 1969 www.nepa.gov/nepa/regs/nepa/nepaeqia.htm

OECD,1992 www.oecd.org

Sadler B. 1994. Environmental Assessment and Development Policy-making, In *Environmental Assessment and Development-An ZAZA*, World Bank Symposium, *in* Goodland R. and Edmundson V. (eds), pp 3-19, the World Bank, Washington DC.

Sadler B (1998b). Institutional Requirements for Strategic Environmental Assessment. Paper to Intergovernmental Forum, organised by the Ministry for the Environment, Christhcurch, New Zealand.

Sadler B and M Baxter, 1997. Taking Stock of SEA, *Environmental Assessment*, Vol 5, Issue 3, PP 14-16, September 1997(Special issue on SEA).

Sadler B and Verheem R, 1996. *Strategic Environmental Assessment: Status, Challenges and Future Directions.* Ministry of Housing, Spatial Planning and the Environment, The Netherlands and the International Study of Effectiveness of Environmental Assessment.

Therivelriki R; Thompson S S; Wilson E, Heaney D, Pritchard D, 1992. *Strategic Environmental Assessment*, Earthscan Publications, London.

Tonk R.A.M.N. and Verheem R.A.A., 1998. Integrating the Environment in Strategic Decision-making: One Concept, Multiple Forms. Paper presented at the annual meeting of the International Association for Impact Assessment (IAIA'98), 19-26 April 1998, Christchurch, New Zealand.

Wathern P (1988) An Introductory Guide to EIA in Wathern P (ed) *Environmental Impact Assessment: Theory and Practice* (pp.1–28) Routledge, London.

WCED, 1987. *Our Common Future,* Report of the Brundtland Commission, Oxford University Press

Wood, C., 1994. Evaluating impact assessment. *Integrated Environmental Management* **30**, p. 12

UNCED, Rio de Janeiro 1992, www.un.org

UNCED,1992. www.un.org/geninfo/bp/enviro.html

UNEP, 1980. www.unep.org

SUGGESTED READINGS

ADB, 1990. Environmental Guidelines for Selected Industrial and Power Development Projects. Office of the Environment, Asian Development Bank, Bangkok.

Ahmad, Y.J. and Swamy, G.K., 1985. Guidelines to Environmental Impact Assessment in Developing Countries. Hodder & Stoughton, London.

Canter, L.W. 1977. *Environmental Impact Assessment*. McGraw Hill, New York.

Clark, B.D., Chapman, K., Bisset, R., Wathern, P. Barret, M. 1981. A Manual for the Assessment of Major Developments. HMSO, London.

Commission of European Communities. 1985. On the Assessment of Effects of Certain Public and Private Projects on the Environment. Official Journal, L175, 28.5.85.

Holling, C.S. 1987. Adaptive Environmental Assessment & Management. John Wiley & Sons, New York.

Htun, N. 1988. The EIA Process in Asia and Pacific Region. *In:* P. Wathern (ed). Environmental Impact Assessment: Theory and Practice. Unwin Hyman, London.

Kulkarni, V.S., Kaul, S.N. and Trivedi, R.K., 2001. *Handbook of EIA*. Scientific Publishers, India.

Lohani, B., Evans, J.W., Ludwig, H., Everitt, R.R., Carpenter, R A. Tu, S.L. 1997. Environmental Impact Assessment for Developing Countries in Asia. Volume 1: Overview. ADB.

Ministry of Environment and Forests. 2000. EIA Manual.

Munn, R.E., 1979. Environmental Impact Assessment: Principles of Procedures. International Council of Scientific Unions, Scientific Committee on Problems of Environment, Report 5 (SCOPE 5)

Rau, J.G. and Wotten, D.C., 1980. Environmental Impact Analysis Hand Book. McGraw-Hill, New York.

Ramachandra, T.V., Subhash Chandran, M.D., Sreekantha, D M, Rao, G.R., and Ali S., 2004. Cumulative Impact Assessment in the Sharavathi River Basin. *International Journal of Environment and Development*, 1 (1): 113-135.

UNCED. 1992. *Rio Declaration*. United Nations Conference on Environment and Development, Rio de Janeiro.

UNEP. 1980. Environmental Assessment Statement—A Test Model Presentation. UNEP Regional Office and UN Asian and Pacific Development Institute, Bangkok.

UNEP. 1980. Guidelines for Assessing Industrial Environmental Impact and Environmental Criteria for the Siting of Industry, UNEP Industry and Environment Guidelines Series, Vol. 1.

UNEP. 1987. Environmental Law Guidelines and Principles: Environmental Impact Assessment, Nairobi.

UNEP. 1988. Environmental Impact Assessment: Basic Procedures for Developing Countries, Nairobi.

Wathern, P. 1988. The EIA Directive of the European Community. *In:* P. Wathern ed., Environmental Impact Assessment: Theory and Practice, Unwin Hymann, London.

WCED. 1987. Our Common Future, World Commission on Environment and Development. Oxford University Press, London

World Bank. 1991. Operational Directive 4.01, Environmental Assessment. The World Bank, Washington, D.C.

World Bank. 1991. Environmental Assessment Sourcebook. Washington, D.C. Vol. 1: Policies, Procedures, and Cross-Sectoral Issues. World Bank Technical Paper 139. Vol. 2: Sectoral Guidelines. World Bank Technical Paper 140. Vol. 3: Guidelines for Environmental Assessment of Energy and Industry Projects. World Bank Technical Paper 154.

UNIT 4

EIA Documentation and Processes

STRUCTURE

OVERVIEW

In Unit 3, we discussed environmental impact assessment (EIA) as a preventive environmental management and decision-making tool. We also touched upon EIA report, known as environmental impact statement (EIS). In that context, we said that EIS is an outcome of an EIA study and documents the environmental consequences of a specific development proposal. We also said that its purpose is to facilitate a full accounting of environmental effects, along with economic and social factors, in decisions as to whether a project should proceed, be modified to ameliorate consequences or be abandoned. We will discuss these in detail in the present Unit, i.e., Unit 4.

In this Unit, we will discuss the processes involved in the preparation and review of EIS. We will begin the Unit by discussing the processes involved in

EIA documentation. We will then explain how EIS should facilitate impact predictions, evaluation and mitigation.

We will then discuss the impact the report can have on decision-making. We will also touch upon the importance of involving the problems in EIA reporting in order that it forms acceptability and EIA monitoring and auditing.

LEARNING OBJECTIVES

After completing this Unit, you should be able to:

- discuss EIA process management and EIA stages;
- explain environmental impact predictions, evaluation and mitigation;
- discuss how EIA reports or EIS contribute to decision-making;
- review, monitor and audit EIA reports for decision-making.

4.1 PRELIMINARY STAGES OF EIA

Before we discuss the stages of EIA, let us spend a few minutes on EIA process management.

Environmental impact assessments often deal with major (and sometimes poorly defined) projects, with many wide-ranging and often controversial impacts. They can also involve many participants with very different perspectives on the relative merits and impacts of projects. It is important that the EIA is well managed, and, essentially, the EIA process is a management-intensive process (Glasson et al., 1994).

The EIA process invariably involves an interdisciplinary team approach, and such an approach reflects the normal scope of EIA studies which range from the biophysical to the socio-economic. The team producing the EIS may be one, or a combination, of the proponents in-house, lead external consultants, external sub-consultants and individual specialists.

The team project manager has a pivotal role to play. In addition to personnel and team management skills, the manager should have a broad appreciation of the project type under consideration, knowledge of the relevant processes and impacts subject to EIA, the ability to identify important issues and preferably a substantial expertise in the area of work. The role of a project manager, in the main, consists of the following:

- selecting an appropriate project team;
- managing specialist inputs;
- liaising with the people involved in the process;
- managing the change in the internal and external environment of the project;

- co-ordinating the contributions of the team in the various documentary outputs.

The management team has to co-ordinate the resources, i.e., information, people and equipment, to achieve an EIA study of quality, on time and within its budget.

> **LEARNING ACTIVITY 4.1**
>
> Explain in no more than 50 words, what will happen if the project proponent prepares the EIA report.
>
> **Note:**
>
> a) Write your answer in the space given below.
>
> b) Check your answer with the one given at the end of this book.

In Subsection 3.1.2 (Unit 3), we listed the steps in the EIA process. In what follows in the present Section, we will discuss some of these steps in detail.

The preliminary stages of an EIA consist of the following steps:

- Project screening.
- Scoping.
- Consideration of alternatives.
- Establishment of environmental baseline.
- Impact identification.

In Subsections 4.1.1 to 4.1.5 we will discuss each of these steps.

4.1.1 Project Screening

The number of projects that could be subjected to EIA is potentially very large. Yet, many projects have no substantial or significant environmental impacts. A screening mechanism seeks to focus on those projects with potentially significant adverse environmental impacts or where the impacts are not fully known. Those with little or no impacts are screened out and are allowed to proceed to the normal planning permission/administrative processes without any additional assessment and without additional loss of time and expense.

Screening can be partly determined by the EIA regulations operating in a country at the time of the assessment. In countries like the UK, there are some projects that will always be screened out for full assessment, by virtue of their scale and potential environmental impacts (e.g., a crude oil refinery, a sizeable thermal power station, a special road, etc.). There are many other projects where the screening decision is less clear. In this scenario, two examples of the project may be screened in different ways (one 'in', one 'out' for full assessment) based on a combination of criteria, including project scale, sensitivity of proposed location and expectation of adverse environmental impacts.

Some EIA procedures include an initial EIA study to check on the likely environmental impacts and their significance. An initial study, which is usually a simple checklist against which environmental impacts must be ticked as 'yes', 'maybe' or 'no' must substantiate the declaration. If the responses are primarily 'no', most of the 'yes' and 'maybe' can be mitigated, then the project may be screened out from a full EIA (Glasson et al., 1994).

LEARNING ACTIVITY 4.2

List three aspects that need to be considered in the screening of projects for EIA.

Note:

a) Write your answer in the space given below.

b) Check your answer with the one given at the end of this book.

4.1.2 Scoping

The scope of the EIA depends on the impacts and issues that it addresses. The process of scoping is to determine the key impacts of the project. An initial scoping of possible impacts may help distinguish the key impacts from those thought to be non-significant and unclear. Further studies should examine the impacts in the various categories. In the process, those identified as non-significant are eliminated and those in the uncertain category, but maybe potentially significant, are added to the initial category of other potentially significant impacts. This refining of focus onto the most significant impacts continues throughout the EIA process.

Scoping is carried out generally in discussions among the developer, the competent authority, other relevant agencies and ideally the public. It is often the first stage of negotiations and consultation between a developer and other interested parties. It is an important step in EIA because it enables the limited resources of the team preparing an EIA to be allocated to the best effect, and prevents misunderstanding among the parties concerned about the information required in the report. Scoping can also identify issues that should later be monitored.

Scoping should begin with the identification of individuals, communities, local authorities and statutory consultees likely to be affected by the project.

LEARNING ACTIVITY 4.3

What are the advantages of scoping in the EIA process?

Note:

a) Write your answer in the space given below.

b) Check your answer with the one given at the end of this book.

A good practice is to bring them together in a working group and/or meetings with the developer. Other key issues could include particularly valued environmental attributes; those impacts perceived to be of particular concern to the affected parties; and social, economic, political and environmental issues related to the specific locality. Reference should be made to relevant structure plans, local plans, subject plans and government policies and guidelines. The end result of this process of information collection and negotiation should be the identification of key issues and impacts, an explanation of why other issues are not considered significant and, for each key impact, a defined temporal and spatial boundary within which the impact will be measured (Glasson et al., 1994).

4.1.3 Consideration of Alternatives

The question of options is particularly important to decide whether or not the selected option is really justified. In other words, by considering alternatives, the worth of the option is determined with other credible alternatives, in environmental, functional, economic and social terms. The rigorous comparison of viable alternatives:

- strengthens decision-making;
- strengthens public credibility;
- improves the methodology of the EIA and the integration of the techniques used for relative comparisons.

To help decision-makers, an EIA report must present clear choices on the planning and implementation of the project, and it should make clear the likely results of each option. For instance, to mitigate adverse impacts, the EIA could propose:

- pollution control technology or design features;
- reduction, treatment or disposal of wastes;
- compensation or concessions to affected groups.

It should be obvious that options must be comparable. In other words, they must be developed to an equal level of detail, a level that must permit measurement of impacts, again at an equivalent level.

In reality, this condition can be met only if the project under examination has evolved by examining options. For example, a common failing lies with site selection. The purchase of sites in the effort to force on them a project that is unsuited to the planning environment, or to the physical and foundation conditions, has been a frequent mistake that would have emerged clearly had the siting options been evaluated.

If you are presented with a project in which site selection has not been the result of a credible options study, or in which the industrial process chosen suffers a similar deficiency, you have a problem that can be solved only by persuading the project sponsor to carry out option studies at a later stage. It is probable that different personnel from those associated with the original project development would have to be employed. While comparisons should as far as possible be quantitative, this is not feasible in all cases. For example, a comparison of the social, cultural or visual impacts of project options must be inevitably the outcome of subjective judgements.

At the simplest level of quantitative comparison, there is the quantified volume of earthwork for two optional road routes, or the measured area of forest eliminated, or the number of stream crossings involved. These all will give some primary guidance as to the extent of the impact that will perhaps be sufficient for a preliminary assessment. Each of the impacts would have to be followed further in a formal EIA. For example, while earthwork quantity may give a first order assessment of possible damage, the effect on water table may be related to depth of cutting. The options may have different visual significance, and therefore, the location and effects of soil dumps may also have to be considered. The area of forest removed does not (but should) take into account forest composition effects, or the presence or absence of rare plants.

Note that this step, i.e., consideration of alternatives, is not a requirement in some countries. This is, in other words, more of a voluntary action than of a mandatory requirement in these countries.

LEARNING ACTIVITY 4.4

Do you feel that the procedure for EIA can be defined as a procedure for designing engineering systems? Justify your answer in no more than 50 words.

Note:

a) Write your answer in the space given below.

b) Check your answer with the one given at the end of this book.

4.1.4 Establishing the Environmental Baseline

The establishment of environmental baseline includes both the present and likely future state of the environment, assuming that the project is not undertaken, taking into account changes resulting from natural events and from other human activities (Glasson et al., 1994). For example, the population of a species of fish in a lake may already be declining even before the potential introduction of an industrial project on the lakeshore.

Initial baseline studies may be wide ranging, but comprehensive overviews can be a waste of resources. The studies should focus as quickly as possible on those aspects of the environment that may be significantly affected by the project, either directly or indirectly. The rationale for the choice of focus should be explained with reference to the nature of the project and to initial scoping and consultation exercises. Although the studies would normally take the various environmental elements separately, it is also important to understand the interaction between them and the functional relationship involved–flora, for example, will be affected by air and water quality and fauna will be affected by flora. This will facilitate prediction.

Baseline studies can be presented in the EIS in a variety of ways. These often involve either a brief overview of both the physical and socio-economic environments of the area of study, following the project description, with the detailed focus studies in subsequent impact chapters (e.g., air quality, geology, employment). Alternatively, a more comprehensive set of detailed studies is presented at an early stage, providing a point of reference for future and often brief impact chapters. A framework for analysing each baseline sub-element is given in Table 4.1.

The final step, i.e., impact identification is the most important and explains the various methodologies used in EIA. Before we discuss this step, let us do Learning Activity 4.5.

LEARNING ACTIVITY 4.5

List the months that are important for baseline data collection for establishing ambient air quality, surface water quality and groundwater quality.

Note:

a) Write your answer in the space given below.

b) Check your answer with the one given at the end of this book.

____

____

____

____

Table 4.1 Framework for Analysing Each Baseline Sub-Element

Sub-Element	*Objectives*	*Required Information/ Specialist(s)*	*Methodology*	*Findings/ Measurements*
Water quality	Protection of human and aquatic life	Existing water quality; possible sources of pollution: runoff, leakage from waste treatment system, surface seepage of pollutants, intrusion of saline or polluted water; capacity of treatment system. Water quality analyst; aquatic biologist; water pollution control engineer; sanitary and civil engineers	Laboratory analyses or field measurement of water quality; pollution indices	Potential for degradation of water quality; safety of potable water.
Surface water	Protection of plant and animal life; water supply for domestic and industrial needs; natural water purification systems; groundwater recharge and discharge; and recreation and aesthetic values.	Location of surface waters—streams, rivers, ponds, lakes, etc.; surface water volume, flow rates, frequency and duration of seasonal variations; 7-day, 10-year low flow; water uses; ecological characteristics; recreation and aesthetic 'uses'. Hydrologist/ecologist.	Measure proximity of site to surface waters; field measurement of volume, rate and direction of water movement; categories of water usage; ecological assessment.	Potential modification of volume, rate and direction of water movement; impact on ecological character; degree and type of water usage.

4.1.5 Impact Identification

Impact identification brings together project characteristics and baseline environmental characteristics with the aim of ensuring that all potentially significant environmental impacts (adverse or favourable) are identified and taken into account in the EIA process. A wide range of methods has been developed for the purpose.

Many of the methods were developed in response to the NEPA and have since been expanded and refined. The simplest involves the use of lists of impacts to ensure that none has been left out. The most complex include the use of interactive computer programmes, networks showing energy flows and schemes to allocate significance weightings to various impacts.

In choosing a method, however, the analyst needs to:

- ensure compliance with regulations;
- provide a comprehensive coverage of a full range of impacts including social, economic and physical;
- distinguish between positive and negative, large and small, long-term and short-term, reversible and irreversible impacts;
- identify secondary, indirect and cumulative impacts as well as direct impacts;
- distinguish between significant and insignificant impacts;
- allow comparison of alternative developmental proposals;
- consider impacts within the constraints of an area's carrying capacity;
- incorporate qualitative as well as quantitative information;
- be easy and economical to use;
- be unbiased and to give consistent results;
- be of use in summarising and presenting impacts in the EIS.

Environmental Impact Assessment Methodologies

There is no single 'best' methodology for environmental impact assessment. Characteristics of a methodology such as the type of impacts or projects covered and the resources required may be virtues in one, but vices in another. Only the user can determine which tools may best fit a specific task. In selecting the most appropriate tools, the following key considerations may be useful:

(i) **Use:** Is the analysis primarily a decision or an information document? (A decision document is vital to determine the best course of action, while an information document is intended primarily to reveal the implications of a single, clearly best choice.) A decision document analysis will generally require a more comprehensive analysis concentrating on interpreting the significance of a broader spectrum of possible impacts.

(ii) **Alternatives:** Are alternatives fundamentally or incrementally different? If differences are fundamental, such as preventing flood damage by levee construction as opposed to flood plain zoning, for example, then impact significance can better be measured against some absolute standard than by direct comparison of alternatives, since impacts will differ in kind as well as size. Fundamentally and incrementally different alternative sets require different levels of analysis to discriminate between alternatives and also require greater degree of quantification.

(iii) **Public involvement:** Does the anticipated role of the public in the analysis involve substantive preparation, token review, or vital review? The first two roles allow the use of more complex techniques such as computer or statistical analysis that might be difficult to explain to a previously uninvolved but highly concerned public. A substantive preparation role will also allow a greater degree of quantification or weighting of impact significance through the direct incorporation of public values.

(iv) **Resources:** How much time, skill, money, data and computer facilities are available? Generally, more quantitative analysis requires more of everything.

(v) **Familiarity:** How familiar is the analyst with both the type of action contemplated and the physical site? Generally, familiarity will improve the validity of a more subjective analysis of impact significance.

(vi) **Issue significance:** How big is the issue in terms of controversy and scope? All other things being equal, the bigger the issue the greater the need for explicitness, quantification, and identification of key issues and the less appropriate is the arbitrary significance weights or specific formulas for trading-off one type of impact (e.g., environmental) against another type (e.g., economic).

(vii) **Administrative constraints:** Are choices limited by the agency procedural or format requirements? Specific agency policy or guidelines may rule out some tools by specifying the range of impacts to be addressed, the need for analysing trade-offs, or the time frame of analysis.

Categorising Methodologies

The various methodologies examined can be divided into the following five types based on the way impacts are identified:

(i) **Ad hoc:** These methodologies provide minimal guidance to impact assessment beyond suggesting broad areas of possible impacts (e.g., impacts on flora and fauna, impacts on lakes, forests, etc.), rather than defining specific parameters to be investigated.

(ii) **Overlays:** These methodologies rely on a set of maps of environmental characteristics (physical, social, economic, aesthetic, etc.) for a project area. These maps are overlaid to produce a composite characterisation of the regional environment. Noting the impacted environmental characteristics lying within the project boundaries identifies impacts.

(iii) **Checklists:** These methodologies present a specific list of environmental parameters to be investigated for possible impacts but do not require the establishment of direct cause-effect links to project activities. They may or may not have guidelines on how parameter data are to be measured and interpreted. The basic categories of checklists are:

- simple checklists (these consist of lists of environmental, economic and social factors, which may be affected by specific project actions) and
- questionnaire checklists (these present a series of questions relating to the impacts of a project. Note that a sample questionnaire checklist is given below). Note that this questionnaire checklist was developed by the US Agency for International Development (1980). The method was designed for the assessment of rural development projects in developing countries. The questions are listed under generic categories such as terrestrial ecosystems and disease vectors. Those assessing projects must attempt to answer all the questions and three answers are possible, depending upon how much is known about a particular impact.

(iv) **Matrices:** These methodologies incorporate a list of project activities in addition to a checklist of potentially impacted environmental characteristics. These two lists are related in a matrix, which identifies cause-effect relationships between specific activities. Matrix methodologies may specify the actions and the corresponding impact on environmental characteristics or may simply list the range of possible actions and characteristics in an open matrix to be completed by the analyst.

(v) **Networks:** These methodologies work from a list of project activities to establish cause-condition-effect-networks. They are an attempt to recognise that a series of impacts may be triggered by a project action. These approaches, generally, define a set of possible networks and allow the user to identify impacts by selecting and tracing out the appropriate project actions.

A Sample Questionnaire Checklist

In what follows, we present a section of a questionnaire checklist dealing with terrestrial ecosystems.

---

a. Are there any terrestrial ecosystems of the types listed below which, by nature of their size, abundance or type, could be classified as significant or unique?

Forest?	Yes	No	Unknown
Savanna?	Yes	No	Unknown
Grassland?	Yes	No	Unknown
Desert?	Yes	No	Unknown

b. Are these ecosystems

Pristine?	Yes	No	Unknown

		Yes	No	Unknown
	Moderately Degraded?	Yes	No	Unknown
	Severely Degraded?	Yes	No	Unknown
c. Are there present trends towards alteration of these ecosystems through cutting, burning, etc. to produce agricultural, industrial, or urban land?		Yes	No	Unknown
d. Does the local population use these ecosystems to obtain non domesticated:				
	Food plants?	Yes	No	Unknown
	Medicinal plants?	Yes	No	Unknown
	Wood products?	Yes	No	Unknown
	Fiber?	Yes	No	Unknown
	Fur?	Yes	No	Unknown
	Food animals?	Yes	No	Unknown
e. Will the project require clearing or alteration of:				
Small areas of land in these ecosystems?		Yes	No	Unknown
Moderate areas of land in these ecosystems?		Yes	No	Unknown
Large areas of land in these ecosystems?		Yes	No	Unknown
f. Does the project rely on any raw materials (wood, fibre) from these ecosystems?		Yes	No	Unknown
g. Will the project decrease use of products from these ecosystems by producing or providing substitute materials?		Yes	No	Unknown
h. Will the project cause increased population growth in the area, bringing about increased stress on these ecosystems?		Yes	No	Unknown

Estimated Impact on Terrestrial Ecosystems...
ND..HA..MA..LA..O..LB..MB..HB

Legend

ND	-	Not determinable	HA	-	High adverse
MA	-	Medium adverse	LA	-	Low adverse
O	-	Low or insignificant	LB	-	Low benefit
MB	-	Medium benefit	HB	-	High benefit

Note that the EIA process is set within an institutional context and the context will vary from country to country. The various steps in the process can be taken in different sequences, and some may even be missing in certain countries.

LEARNING ACTIVITY 4.6

Can you name a project where preliminary EIA should be carried out? Justify your answer in not more than 50 words.

Note:

a) Write your answer in the space given below.

b) Check your answer with the one given at the end of this book.

Having discussed the preliminary stages, let us now focus on the central steps of impact prediction, evaluation and mitigation.

4.2 IMPACT PREDICTION, EVALUATION AND MITIGATION

EIA is all about prediction and is needed at the earliest stages when the project, including alternatives, is being planned and designed, and this continues through to mitigation, monitoring and auditing. Evaluation follows from prediction and involves an assessment of the relative significance of the impacts. The methods of evaluation range from intuitive to the analytical, from qualitative to quantitative, and from formal to informal. Cost benefit analysis, monetary valuation techniques, and multi-criteria/multi-attribute methods, with their scoring and weighting systems, provide a number of ways for the evaluation issues. Mitigation of significant adverse effects involves the measures to avoid, reduce, remedy or compensate for the various impacts associated with projects. We will discuss these in Subsections 4.2.1 to 4.2.3, respectively.

4.2.1 Prediction

Environmental impact is any alteration of environmental conditions or creation of a new set of environmental conditions–adverse or beneficial–caused or induced by the project under consideration. The impact depends on the nature, scale and location of the proposed activity, and it includes the effect on the natural resource base (i.e., the quality of air, water, noise, biological) and socio-economic components of the environment which determine the cost of environmental management.

The impacts can be classified as primary, which can be attributed directly to the project, and secondary, which are indirect changes and typically include the changed patterns of socio-economic activities likely to be stimulated or induced by the proposed activity.

An EIA partly answers the question: What will happen as a result of the project? But, if a preliminary assessment is carried out, it will broadly review the project's effects. Also, scoping helps the decision-makers identify the most important issues. Taking these findings into account and after collecting the baseline environmental data, the full EIA study formally identifies such of the impacts as are to be assessed in detail. The methods used at the identification phase of the study include the following:

(i) Compile a candidate list of key impacts such as changes in air quality, noise levels, wildlife habitats, species diversity, landscape views, social and cultural systems, settlement patterns and employment levels. This data is drawn from as many EIA studies carried out on similar projects as possible.

(ii) Name all the project's *sources* of impacts (e.g., smoke emissions, water consumption, construction jobs) using checklists or questionnaires and then list possible *receptors* in the environment (e.g., crops, communities using the same water for drinking, migrant labourers) by surveying the existing environment and consulting with interested parties. Where the sources affect the receptors, a potential impact is suspected.

After identifying the environmental components likely to be impacted by the proposed activities, the next logical step is impact prediction, which answers the question: What will be the extent of the changes? Prediction scientifically characterises the impact's causes and effects, and its secondary and synergistic consequences for the environment and the local community. It follows an impact within a single environmental parameter (e.g., a toxic liquid effluent) into its subsequent effects in many disciplines (e.g., reduced water quality, adverse impacts on fisheries, economic effects on fishing villages and resulting socio-cultural changes). Prediction draws on the physical, biological, socio-economic and anthropological data and techniques. In quantifying impacts, it may employ various tools such as mathematical models, photomontages, physical models, socio-cultural models, economic models, experiments, expert judgement, etc.

To prevent unnecessary expenses, the sophistication of the prediction methods used should be kept in proportion to the scope of the EIA. For instance, a complex mathematical model of atmospheric dispersion should not be used, if only a small amount of relatively harmless pollutant is emitted. Simpler models are available and are sufficient for the purpose. Also, it is unnecessary to undertake expensive analysis, if they are not required by the decision-makers for whom the EIA is being done.

All prediction techniques, by their nature, involve some degree of uncertainty. So, along with each attempt to quantify an impact, the study team should also quantify the prediction's uncertainty in terms of probabilities or margins of error.

One of the criticisms against EIA is that it does not give the prominence the social and cultural impacts deserve in describing the extent of changes expected to result from a major development project. Socio-cultural impacts should be integrated into every discussion of physical/biological change.

4.2.2 Evaluation

This step evaluates the predicted adverse impacts to determine whether they are significant enough to warrant mitigation. In other words, evaluation attempts to find an answer for the question: Do the changes matter? The judgement of significance can be based on one or more of the following:

- Comparison with laws, regulations or accepted standards.
- Consultation with the relevant decision-makers.
- Reference to pre-set criteria such as protected sites, features or species.
- Consistency with government policy objectives.
- Acceptability to the local community or the general public.

The Scaling-weighting Technique

Scaling of the effects addresses issues of magnitude and is based on a numerical system in which the highest number represents a very good effect and the lowest number represents a very adverse effect. The mid-point indicates an average effect, or a neutral one. Scaling can be used either alone to determine some composite score for magnitude or with a weighting scheme to incorporate considerations of importance or significance. With weighting, values are assigned to elements of the environment (or of a proposal) based on their relative importance or significance. The end result can be either a total score or a group of scores representing various sectors of the environment.

There are several methods that incorporate some means of *scaling* the effects and *weighting* the relative importance of various elements, thus resulting in a mathematical means of combining magnitude and significance for an overall evaluation of the project impact. Technically, these methods fall into the category of checklists, but, in reality, they go far beyond what checklists provide.

A combination of scaling and weighting is particularly useful because it combines measures of magnitude and significance in a highly organised format, and, therefore, is valuable for analysing alternatives. This technique incorporates a large amount of data that may not otherwise be readily comparable. Thus, it provides a means of synthesising a great deal of information while still retaining measures of magnitude and significance.

Judgement is inherent in this approach, particularly when weights or significance are assigned to individual components or elements. However, for objective evaluation, it is necessary to define significance very carefully. In addition, when scaling the effects, it is necessary to be explicit about what the values mean. Scaling works best when the scales relate to real, measurable changes, as in water quality or employment levels. Given appropriately detailed documentation, the scaling and weighting values decided upon by different people should not vary much. An example of a scaling-weighting technique for site selection is given in Fig. 4.1.

Cost-benefit Analysis

The scope of approaches such as financial-appraisal, cost-minimisation and cost-effectiveness methods, is limited as they consider only a subsection of the relevant population or only a subsection of the full range of consequences of a plan or project. For example, financial appraisal is limited to a narrow concern, usually of the developer, with the stream of financial costs and returns associated with an investment. Cost effectiveness involves selecting an option that achieves a goal at the least cost. The cost effectiveness approach is more problematic where there are a number of goals and where some actions achieve certain goals more than others.

In contrast, a cost-benefit analysis is more comprehensive in scope. It takes a long view of projects (farther as well as nearer to the future) and a wide view (in the sense, allowing for side effects). It is based in welfare economics, and seeks

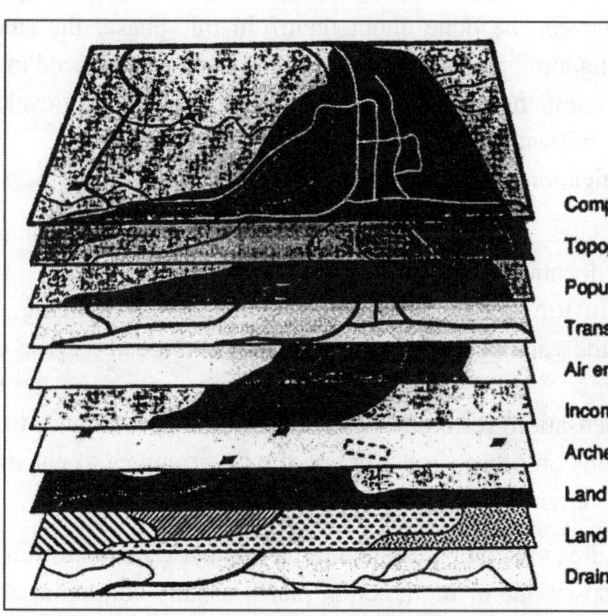

Figure 4.1 Scaling-Weighting Technique: Example of an Overlay
Source: Dixon and Montz, 1993

to include all the relevant costs and benefits to evaluate the net social benefit of a project.

A cost-benefit analysis consists of several stages such as project definition, identification and enumeration of costs and benefits, evaluation of costs and benefits, discounting and presentation of results, etc. Note that, some of these stages are similar to those in EIA. The basic evaluation principle is to measure in monetary terms—as money is the common measure of value and monetary values are best understood by the community and decision-makers—and then to reduce all costs and benefits to the same capital on annual basis. Future annual flow of costs and benefits are usually discounted to a net present value. A range of interest rates may also be used to show the sensitivity of the analysis to changes. If the net social benefit minus cost is positive, then there may be a presumption in favour of a project. However, the final outcome may not be always clear. The presentation of results should distinguish between tangible and intangible costs and benefits, as relevant, allowing the decision-maker to consider the trade-offs involved in the choice of one option or another.

4.2.3 Mitigation

If the answer to the evaluative question (i.e., Do the changes matter?) is in the affirmative, i.e., the changes do matter, then the EIA proceeds to find an answer to the question: What can be done about them? In this phase, the study team formally analyses mitigation, and a wide range of measures is advanced to prevent, reduce, remedy or compensate for each of the adverse impacts evaluated as significant. Since all mitigation measures cost, it must be quantified.

The possible mitigation measures include:

- Changing project sites, routes, processes, raw materials, operating methods, disposal routes or locations, timing, or engineering designs.
- Introducing pollution controls, waste treatment, monitoring, phased implementation, landscaping, personnel training, special social services or public education.
- Offering (as compensation) restoration of damaged resources, money to affected persons, concessions on other issues, or off-site programmes to enhance some other aspect of the environment or quality of life for the community.

Once these measures are compared, and trade-offs between them weighed, the EIA study team proposes one or more action plans, usually combining a number of measures. The action plan may include technical control measures, an integrated management scheme (for a major project), monitoring, contingency plans, operating practices, project scheduling, or even joint-management (with affected groups). The study team should explicitly analyse the implications of adopting different alternatives to help make the choices clearer for decision-makers. Several analytical techniques including the following are available for this purpose:

- Cost/benefit analysis, in which all quantifiable factors are converted to monetary values and actions are assessed for their effect on project costs and benefits. (Note, however, that the unquantifiable and qualitative aspects can be equally important, and often need to be taken into account in the decision-making process.)
- Explaining what course of action would follow from various broad value judgements (e.g., that social impacts are more important than resources).
- A simple matrix of environmental parameters versus mitigation measures, containing brief descriptions of the effects of each measure.
- Pair-wise comparisons, whereby the effects of an action are briefly compared with the effects of each of the alternative actions, one pair at a time.

The mitigation plan should be supplemented with an environmental management plan (EMP) to guide the proponent towards environmental improvements. The

EMP is a crucial input to monitoring the clearance conditions and, therefore, the details of monitoring should be included in the EMP.

Having discussed impact prediction, evaluation and mitigation, we will next explain another important stage of EIA, i.e., impact on decisions.

4.3 IMPACT ON DECISIONS

The primary purpose of EIA studies, as mentioned earlier, is to aid decision-making and serve as an instrument for sustainable development. Decisions can be made at various levels. A national department of energy may take decisions on the direction of energy policy (e.g., the mix of coal-fired versus gas-fired electricity generation) and on the resultant sector programmes (e.g., a 10-year programme to develop a combined cycle gas turbine (CCGT) plant). Various combinations of stakeholders (e.g., private sector, public sector) can also take decisions at various stages in the planning and development process.

Figure 4.2 represents a sound integrated decision-making process.

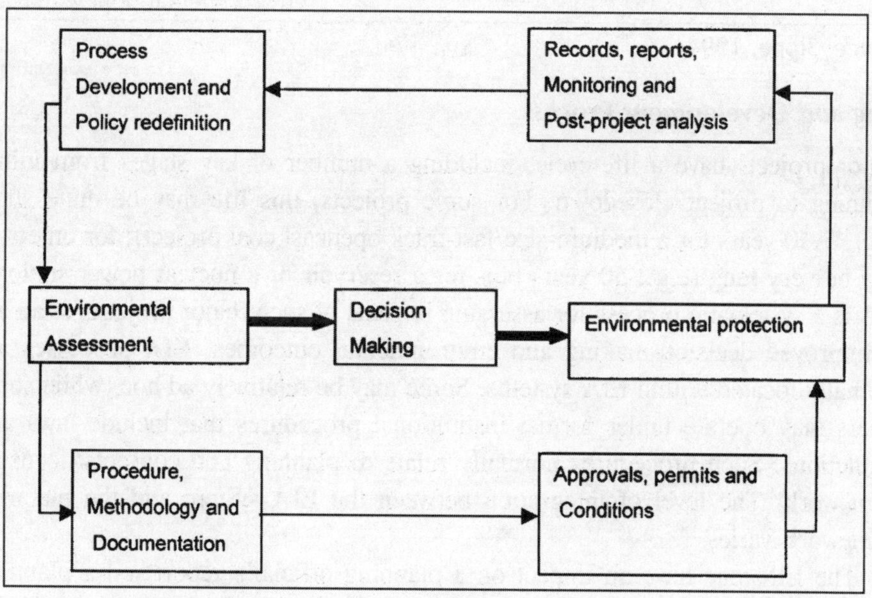

Figure 4.2 Environmental Considerations in Decision-Making

Table 4.2 illustrates both socio-economic and bio-physical environmental categories with examples of negotiable and non-negotiable impacts.

Table 4.2 Judging Environmental Acceptability

	Non-negotiable impacts	*Negotiable impacts*
Ecological (physical and biological components)	Degrades essential life support systems. Degrades conservation. Estate. Adversely affects ecological integrity. Loss of biodiversity.	No degradation beyond carrying capacity. No degradation of productive systems. Wise use of natural resources.
Social (humans as individuals or in social groupings)	Loss of human life. Reduces public health and safety unacceptably. Unreasonably degrades quality of life where people live.	Community benefits and costs and where they are borne. Reasonable apportionment of costs and benefits. Reasonable apportionment of intergenerational equity. Compatibility with defined environmental policy goals.

Source: Sippe, 1994

Decisions, Planning and Development Process

Major projects have a life-cycle, including a number of key stages from initial planning to project closedown. For some projects, this life may be quite short (e.g., 5–10 years for a medium-size/fast-track opencast coal project); for others, it may be very long (e.g., 50 years plus for a reservoir or a nuclear power station). EIA is a systematic process for assessing impacts of such major projects as an aid to improved decision-making and environmental outcomes. EIA processes are normally located within EIA systems. Some may be relatively ad hoc, while some others may operate under formal institutional procedures that include laws and regulations. Such procedures normally relate to planning and consents/licensing framework. The level of integration between the EIA system and the planning framework varies.

The EIS may have an impact on a planning official's report, on a planning committee's decision, and on pre-and post-submission project modifications. Decisions are rarely a simple 'yes' or 'no', and are more likely to be conditional. For example, project support is pledged subject to certain modifications and/or associated measures. Whilst this is important, and EIA can have a key role to play at this stage, decision-making occurs at many other stages and is undertaken by various parties. Decisions, for example, are to be made on whether EIA is needed at all (the screening stage); the scope of the EIA; the alternatives under

to communicate information and to seek advice on the basis that the advice, if relevant, will be acted upon.

- Public involvement is an area of the EIA reporting process that can be politically sensitive. Certain interests, even project sponsors, may seek to exercise control and influence on the process.

The process you devise, hopefully through wide consultation, will be inevitably project-specific because the sponsors, the people directly affected and the general public interest, will be different for each project. There will not be one public so far as a specific project is concerned. There may, in fact, be several *publics* as under:

- The experts within the community, the scientific organisations, the expert government agencies, university departments and expert professional groups.
- Local authorities, citizen groups and NGOs.
- The stakeholders, i.e., those with a direct interest in, or who are directly affected by, the project.
- Societies, cultural groups and individual citizens interested in, or affected by, the project.
- The general community.

Identifying the publics in relation to your particular project, and perhaps consulting to establish the level of information and involvement they would like to have, will do a lot for your credibility and the ultimate success of your EIA report. As well, it will help ensure that knowledge of the intention to carry out an EIA report is disseminated at a vital time–before it starts. One of the most certain routes to damaged credibility and unnecessary objections is the discovery by affected citizens that their opportunity for intervention has been pre-empted. This can be expressed as the first of the following general rules governing participation:

- make sure that all the identified publics have been advised before the EIA;
- state about the project, its objectives, programmes, the proposed public involvement process, and the anticipated documentation;
- information must be communicated, not just provided;
- language, format, illustration and media vehicle is appropriate to a particular public.

You will find that you have a number of options (and opportunities) for the type of consultation best suited to a particular public or phase of your EIA Report. These include newspaper advertisements, reports of journalists, public meetings, cottage (or small scale) meetings with specific interest or stakeholder groups, TV and radio, newsletters and progress reports.

In summary, the benefits of a quality public involvement programme are immense. The greatest benefit will be trust and confidence. Your conclusions, even if unpalatable, will be more likely to get acceptance if the public develops confidence in your work, methods and your objectivity and professionalism.

4.4.2 Requisites for a Good EIS

An EIS should be comprehensive. It must, at the minimum, fulfil the requirements of the relevant EIA legislation concerning contents. A good EIS will also go further than the minimum requirements, if other significant impacts are identified. Most EISs are broadly organised into the following sections:

(i) a non-technical summary;
(ii) a discussion of relevant methods and issues;
(iii) a description of the project and of environmental baseline conditions;
(iv) a discussion of the project's likely environmental impacts (which may include the discussion of baseline environmental and predicted environmental impacts, proposed mitigation measures and residual impacts).

Ideally an EIS should also include the main alternatives that were considered, and proposals for monitoring. It should emphasise key points. These should have been identified during the scoping exercise, but additional issues may arise during the course of the EIA. The EIS should set the context. It should be kept as brief as possible while still presenting the necessary information. The main text should include all relevant discussion of impacts, and appendices should present only additional data and documentation.

EIAs are, indirectly, public relation exercises and can be seen as publicity documents for developers. Good presentation can convey a concern for the environment, a rigorous approach to the impact analysis and a positive attitude to the public. Bad presentation, in turn, suggests a lack of care and perhaps a lack of financial backing. Similarly, good presentation can help to convey information clearly, whereas bad presentation can negatively affect even a well-organised EIS (Glasson et al., 1994).

4.4.3 Review of EIS

An environmental impact assessment (EIA) report, i.e., EIS, is normally reviewed once it has been completed, before being submitted to the decision-making authority. The main purpose of the review is to verify that the information and conclusions presented in the EIA report are sufficient for informed decision-making.

A proper review and evaluation of EIS requires a triple-A test of *appropriateness* (i.e., coverage of key issues and impacts), *adequacy* (i.e., impact analysis), and *action ability* (i.e., does the report provide the basis for informed decision making?).

Essentially, an EIA report satisfies the following three criteria:

(i) Completeness and conformance with the TOR (terms of reference) for the EIA.

(ii) Accuracy and veracity as defined by general acceptable scientific criteria (e.g., quality assurance and quality control procedures for analysis of sampling data) and use of acceptable methods for the assessment of environmental impacts.

(iii) Clear description of environmental impacts, recommended mitigation measures, environmental monitoring plan and environmental management plan.

Now, consider the following checklist for reviewing an EIA report:

(i) Executive summary

- An adequate summary of the significant findings of the EIA report.
- A sufficiently detailed description of how significant environmental issues will be resolved.
- A presentation of the study's conclusions.
- Effective, simple, visual presentations of the type and magnitude of the impacts.

(ii) Introduction

- Description of project rationale.
- Consideration and evaluation of project alternatives.
- Methods used to identify, predict and assess impacts.

(iii) Project description

- A listing of project activities that are likely to cause significant impacts to environmental resources.
- A listing of mitigation measures to be incorporated into the project.
- The location, scale and scheduling of activities.
- Potential accident or hazard scenarios covered in the risk assessment are based on the characteristics of the project and the history of accidents at similar types of facilities.
- Information should match the expected operations according to the feasibility study.

(iv) Description of the baseline environment

- A description of the environmental components that may be significantly affected by the project.
- An explanation of the derivation of environmental indicators chosen to represent environmental components.
- Base maps for spatial data.
- Baseline values, or some other appropriate form of quantitative and qualitative information, for environmental components/resources that may be affected either directly or indirectly by project activities.

(v) Prediction/assessment of impacts

- A description of the major issues.
- Documentation of the cause and effect relationships between planned project activities and the environmental components.
- Identification of secondary or higher order effects with clearly defined pathways of impacts from higher order effects.
- Impact prediction includes a number of stated assumptions that affect the predicted impacts, their probability of occurrence and degree of impact.
- Methods used to predict impacts.
- An assessment of the significance of predicted impacts, methods or approaches to assigning impact significance.
- Justification for the choice of methods used to predict environmental impacts.

(vi) Development of mitigation measures

- A description of all the environmental protection measures considered to mitigate or offset damaging environmental impacts from project activities.
- A description of the costs and benefits for each recommended environmental protection option developed to resolve a significant environmental issue, as well as a comparison of each option to the other options.
- Appropriateness and cost effectiveness of environmental protection measures.
- A description of the technology used in each environmental protection measure, including information regarding its prior effective use, the range of environmental conditions under which it is effective, and the level of skill required to operate or maintain the technology.
- A time schedule for implementation of the environmental protection measures, showing that they will be in use before the project impacts are felt.

- A drawing or table that illustrates how the mitigation measures address the significant environmental issues.

(vii) Public consultation and participation

- Description of strategy and approach.
- Chronology of individuals and groups consulted.
- Descriptions of methods used to consult with public.
- Summary of information obtained during consultations and how it was used in preparation of the EIA report.
- A sampling programme design (frequency, intensity) sufficient to provide the information necessary to answer questions.
- Analytical system quality assurance and quality controls are effective.
- Information for reporting monitoring results.

(viii) Environmental management plan and institutional arrangements

- Detailed descriptions of the environmental protection measures, the monitoring programme, and follow-up activities.
- Monitoring requirements.
- Allocation of responsibilities and tasks.
- Staffing requirements.
- Budgets and schedules.
- Administrative arrangements.
- Administrative mechanisms for enforcement and taking corrective action are in place.

(ix) Summary and conclusions

- Net benefits which justify the project.
- Explanation of how adverse effects have been mitigated.
- Explanation of use or destruction of any irreplaceable components.
- Provisions for follow-up surveillance and monitoring.

Uncertainties can arise in any decision, and EIA is no different. Monitoring and auditing in EIA enables the environmental authority to reconsider its original decision and also whether there is a need for some form of additional action. Let us discuss these issues in Section 4.5. But, first, let us work out Learning Activity 4.7.

> **LEARNING ACTIVITY 4.7**
>
> List three benefits of public participation in EIA process.
>
> **Note:**
>
> a) Write your answer in the space given below.
>
> b) Check your answer with the one given at the end of this book.
>
> ______________________________
>
> ______________________________
>
> ______________________________
>
> ______________________________

4.5 EIA MONITORING AND AUDITING

The issues in EIA monitoring and auditing appear to shift from purely technical and scientific to management aspects. Key issues in EIA monitoring and auditing, thus, are the accuracy of impact prediction and the quality of the EIS. EISs are generally expected to contain testable hypotheses, and monitoring and auditing focussed on compliance. Of late, however, the focus is on project implementation, mitigation, activity management and communication. Nevertheless, EIA monitoring and auditing are less developed than the pre-decision elements of the EIA process. This does not mean that there is no post-decision monitoring and evaluation—in many jurisdictions, as compliance monitoring and enforcement of permits are elements of the environmental permit system.

Table 4.3 gives the relationship between monitoring and auditing objectives and uncertainties in EIA.

Table 4.3 Relationship Between Monitoring and Auditing Objectives and Uncertainties

Objectives of EIA monitoring and auditing	*Uncertainties about*
Controlling	Correct implementation and operation (permit and general environmental standards). Effects on the environment.
Scientific knowledge	Dose-effect relationships (causal mechanisms). Accuracy of prediction methods.
Acceptance and legitimisation	Public concern/debate about the effects of the activity. Public acceptance with respect to the overall plan of which the project forms a part. Public preferences with respect to valuation of impacts.
Integration with other information	None (the objective relates to efficient governance).

4.5.1 Monitoring

A monitoring plan should specify for each indicator the measurement technique, frequency, locations, responsibilities, reporting, data storage method, etc. These choices depend upon the type of activity, the indicators selected for monitoring, the timescale of impacts and the characteristics of the area, particularly of dispersion media and sensitive targets. Monitoring of environmental pollution by point sources is illustrated in Table 4.4:

Table 4.4 Environmental Impact and their Monitorable Characteristics

Environmental impact	*Monitorable characteristics*
Contribution to green house effect	Energy use by type
Contribution to stratospheric ozone depletion	Emission of ozone depleting gases
Acidification	Emission of NO_x, SO_2
	Impact on ecosystems
	Impact on buildings, artificial constructions
Ambient air pollution	Emission of toxic gases
	Ambient air quality
Noise and vibration	Noise emission
	Ambient air quality
Soil pollution	Emission to the soil
	Quality of the soil
	Quality of groundwater near the pollution source
	Quality of groundwater near water intake
Surface-water pollution	Emissions to surface water or sewage system
	Quality of surface water

Environmental pressures may be monitored by adopting environmental management and auditing systems, which consist of measures such as procedures for handling of wastes, prevention of spillage, etc., performance checking and maintenance of protection measures and facilities (such as sampling of groundwater beneath secure disposal sites), registration of generated waste and its method of disposal and registration of material flows in order to discover leaks. These systems maybe specified in the permit conditions.

The activity manager might go further and certify his or her environmental management system under the ISO 14000 series. This type of indicator will give information about the mitigation performance of the operator, and normally emissions will be known to be at an acceptable level. According to the best available technology (BAT) principle, the activity manager should reduce emissions to a minimum, although the latter does not necessarily have to be defined in quantitative terms.

4.5.2 Auditing

Monitoring is only useful if the observations are collated, evaluated and formally reported and if required actions are taken. Auditing implies that the monitoring results are compared with standards or criteria of acceptability and that, subsequently, a value judgement is made in order to decide whether further action is needed.

With reference to the primary aim of control to secure environmental protection, the original consent decision is important in providing the audit reference point. For appraising the monitoring results, information in the EIS may also be relevant (e.g., impacts as predicted and the described reference situation). Other relevant criteria may include legal standards, regulations and policies, new investigation and new insights. Often, the acceptability of monitoring results cannot be defined fully in advance. The audit will be most effective, if the results are made public and if the component authorities have some obligation to report publicly.

On the basis of the audit, the environmental authority can reconsider its original decision and it may also consider whether there is a need for some form of additional action. Remedial actions can be necessary, if the actual impacts exceed the expected or acceptable impacts, or if the original terms and conditions need to be reviewed. Depending on the significance of unacceptable impacts, actions may range from doing nothing, adjusting the permit granted, taking preventive, mitigation or compensation measures, through to stopping the activity (Handbook of Environmental Impact Assessment, Vol. 1).

SUMMARY

In this Unit, we discussed the processes involved in EIA documentation–screening, scoping, consideration of alternatives, baseline establishment and impact identification. Following this discussion, we explained the role of EIS in impact prediction, evaluation and mitigation. Being predominantly a decision-making tool, EIA, we said, must contribute to informed decision-making, and in that context explained how EIS facilitates decisions. We also discussed the importance of involving the public and cautioned that without which the report would not receive the desired level of acceptability. We then mentioned some of the prerequisites for an effective and useful EIS and provided a checklist to review an EIA report. We closed the Unit by explaining the processes involved in, and the importance of, EIA monitoring and auditing.

REFERENCES

Dixon J E and Montz B, 1993. *'From law of practice: EIA in New Zealand"* Environmental Impact. Assessment Review 13, PP 89-108. Sitarz D (ed.), 1993 *Agenda 21: The ...*

Glasson, J., Therivel, R., and Chadwick, A. 1994. Introduction to Environmental Impact Assessment. UCL Press Ltd, London.

Sippe 1994 Sippe, R. 1994. Policy and EA in Western Australia: Objectives, Options, Operations and Outcomes. The Hague: Paper to the International Workshop on Policy EA.

SUGGESTED READINGS

Becker, H.A., 1991. The Future of Impact Assessment in the European Communities and their Member Countries. *Impact Assessment Bulletin,* **9,** No. 4.

Biswas, A.K. and Qu Geping (Chu, Ko-Ping) (eds)., 1987. Environmental Impact Assessment for Developing Countries. Tycooly International, Oxford.

Biswas, A.K. and Agarwala, S.B.C., 1994. *Environmental Impact Assessment*. Butterworth-Heinmann, Oxford.

Canelas, L.D., 1991. Implementation of EEC Directive in Portugal: A Case Study. *Impact Assessment Bulletin,* **9,** No. 4.

Canter, L.W., 1977. *Environmental Impact Assessment*. McGraw-Hill, New York.

Cleark, B., Gilad, A., Bissett, R. Tomlinson, P. (eds). 1984. Perspectives on Environmental Impact Assessment. Reidel, Dordrecht.

Colombo, A.G. (ed). 1992. *Environmental Impact Assessment*. Kluwer Academic Publishers, Dordercht, for the Commission for the European Communities.

Dixon, J.E. and Montz, B.E. 1992. Resource kit prepared for seminar on Mananging Environmental Effects. Centre for Environmental and Resource Studies, University of Waikato, Hamilton, N.Z.

Dixon, J.E. and Montz, B.E. 1993. Assessing Environmental Effects: A guide for practitioners. Occasional paper series, Paper No. 1, Centre for Environmental and Resource Studies, University of Waikato, Hamilton, N.Z.

Devuyst, D. 1993. Environmental Impact Assessment. *In:* Environmental Management, Volume 3, Instruments for Implementation, B. Nath, L. Hens and D. Devuyst (eds). VUB University Press, Brussels.

Economic Commisison for Europe. 1987. Application of Environmental Impact Assessment, Highway and Dams. Environmental Series 1. United Nations, New York.

Ellis, D.V. 1989. Environments at Risk: Case Histories of Impact Assessment, Springer-Verlag, Berlin, 322 pp.

Erickson, P.A. 1994. A practical guide to environmental impact assessment. Academic Press, San Diego.

Fortlage, C.A. 1990. Environmental Assessment: A Practical Guide. Gower, Aldershot.

Gilpin, A. 1994. Environmental Impact Assessment (EIA): Cutting Edge for the 21st Century. Cambridge University Press, Cambridge.

Glasson, J., Therivel, R. and Chadwick, A. 1994. Introduction to environmental impact assessment, principles and procedures, process, practice and prospects. UCL Press, London.

Govorushko, S.M. 1991. Environmental Impact Assessment in the USSR: Current situation. *Impact Assessment Bulletin,* **9,** No. 3.

Hill, R.C., Archer, F.M. and Webley, W.T.A. 1990. Conflict over Change in Land Tenure in the Reserves of Namaqualand, South Africa. *Impact Assessment Bulletin,* **8,** Nos. 1 and 2.

Krta Ltd in Association with Fraser Thomas Partners. 1989. Beachlands-Maraetaind, Sewage Feasibility Study, Auckland.

Institution of Professional Engineers of New Zealand. 1985. The IPENZ Environmental Code.

McHarg, I.L. 1969. *Design with Nature*. The Natural History Press, Garden City, New York. (There are also other publishers.)

Mock, J. F. and Bolton, P. 1994. The ICID Environmental Checklist—To Identify Environmental Effects of Irrigation, Drainage and Flood Control Projects. Overseas Development Unit, Hydraulics Research, Wallingford.

Morris, P. and Therivel, R. 1994. Methods of Environmental Impact Assessment. UCL Press, London.

New South Wales. 1980. Supplement to the Government Gazette, Friday 28 August, 1980.

Nijkamp, P. and Blass, E. 1994. Impact Assessment and Evaluation in Transportation Planning. Kluwer.

OECD, 1994. Environmental Impact Assessment of Roads. OECD, Paris.

Pearce, D., Barbier, E. and Markandya, A. 1990. Sustainable Development, Economics.

Porter, C.F. 1985. Environmental Impact Assessment, A Practical Guide. University of Queensland Press, St. Lucia.

Ramachandra, T.V., Subhash Chandran, M.D., Sreekantha, D M, Rao, G.R. Ali S. 2004. Cumulative Impact Assessment in the Sharavathi River Basin. *International Journal of Environment and Development*, **1 (1):** 113-135.

Thomas, I. 1987. Environmental Impact Assessment: Australian Perspectives and Practice. Graduate School of Environmental Science, Monash University, Melbourne.

Wathern, P. (ed). 1988. Environmental Impact Assessment, Theory and Practice. Unwin Hyman, London.

World Bank Environmental Department. 1991. Environmental Assessment Sourcebook, 2 volumes. World Bank Technical Papers No. 139 and 140, Washington, D.C.

World Commission on Environment and Development. 1987. *Our Common Future*. Oxford University Press, Oxford.

UNIT 5
Environmental Auditing

STRUCTURE

OVERVIEW

In Unit 4, we discussed the preparation and review of an EIA report, and in the present Unit (i.e., Unit 5), we will discuss another tool for environmental management—environmental auditing. As environmental legislations are becoming increasingly stringent across the world, industries are required to demonstrate improvements in environmental performance for their survival and growth. In the wake of environmental regulations that proliferated in the 1970s and 1980s, environmental audit was born as an environmental management tool for industries and other organisations seeking to monitor their compliance performance on environmental matters. The main purpose of environmental audit is to identify, assess and address environmental concerns in order to establish environmental quality assurance.

That said, in the first three Sections of this Unit, we will discuss the basics of environmental auditing (EA) by explaining the objectives, scope and types of EA as well as the methodologies, stages and procedures involved in carrying out an auditing. We will also discuss waste, liability, industrial and EMS audits.

LEARNING OBJECTIVES

After completing this Unit, you should be able to:

- discuss the objectives of an environmental audit;
- describe the basic steps and elements of an environmental audit;
- explain waste audits and pollution prevention assessments;
- discuss liability, industrial and EMS audits;
- discuss the potential of environmental audit as a tool for environmental management to identify, assess and address environmental concerns;
- conduct/co-ordinate an environmental audit and critically evaluate its outcomes.

5.1 INTRODUCTION TO ENVIRONMENTAL AUDITING (EA)

An environmental audit (EA) is a systematic, independent internal review to check whether the results of environmental work tally with the targets. It studies whether the methods or means used to achieve the goals or ends are effective. EA involves studying documents and reports, interviewing key people in the organisation, etc., to assess the level of deviations between targets and results.

It is defined as a systematic and documented verification process of objectively obtaining and evaluating evidence to determine whether an organisation's EMS conforms with audit criteria set by the organisation, and for communicating the results of this process to management (ISO 14001).

Environmental audits are being used as a tool and an aid to test the effectiveness of environmental efforts at local level. They can be carried out for a number of reasons including the following:

- To verify compliance.
- To review implementation of policies.
- To identify liabilities.
- To review management systems.
- To identify needs, strengths and weaknesses.

- To assess environmental performance.
- To promote environmental awareness.

5.1.1 Objectives and Scope

The objectives of an environmental audit are to evaluate the efficiency and efficacy of resource utilisation (i.e., people, machines and materials), to identify the areas of risk, environmental liabilities, weakness in management systems and problems in complying with regulatory requirements and to ensure the control on waste/pollutant generation.

The areas an environmental audit deal with can be categorised as under:

- **Design specification and layout:** While setting up an industry, adequate provisions are made in the design specification and layout to augment the production capacity but corresponding provisions to meet the environmental criteria are often overlooked. Adequate provisions are, therefore, necessary to upgrade pollution control measures to meet the future environmental standards that are getting stringent day by day. The audit will help in identifying specific areas of concern to meet the future requirements of environmental measures.
- **Resource management:** The resources include air, water, energy and other raw materials. The audit will provide data to the management on the efficient use of the resources per unit production, and, thereby, help reduce resource consumption and waste minimisation.
- **Pollution control systems and procedures:** The audit helps ensure that the systems and procedures governing the environmental activities/operations of pollution control equipments are rightly followed and determine the efficiency of the system in identifying conditions and inviting corrective actions in a timely and effective manner.
- **Emergency plans and response/safety system:** As the emergency plans more often than not remain in the safe custody of senior management, staff may not have immediate access to the right action during an emergency. The problem becomes acute when new persons are employed/deployed. The review of the emergency response system will ensure adequate knowledge, alertness and readiness of the staff concerned to effectively face an emergency.
- **Medical and health facilities/industrial hygiene and occupational health:** The productive element of an industry is dependent on the health of its human resources. The primary facilities to suit the occupational needs of the industry are, therefore, vital. Audit in this regard will provide an insight into the actual requirements to warn suitable orientation of existing facilities.
- **Confirmation to regulatory requirement:** The regulatory mechanism of environmental compliance is gradually becoming more and more comprehensive.

New regulations and standards are being stipulated at such a pace that they render the existing systems archaic. Factory managers may not be fully aware of the latest requirements and this will make the top management/owners vulnerable for prosecution under various environmental acts. An audit helps compare the existing status with the stipulation and standards prescribed by various agencies and ensure compliance.

Scope

The scope of an environmental audit can fall under either of the following two options:

(i) To limit scope of the audit to an assessment only of the degree of compliance with policies, requirements-type documents and procedures.
(ii) To assess the adequacy of the policies, requirements—type documents and procedures to begin with and, given their adequacies, to assess the degree to which compliance with these documents is achieved.

The argument for limiting the audit scope only to the assessment of compliance is that the policies, requirements-type documents, and procedures have been developed and approved by the key affected technical and managerial personnel. The audit objectives, however, should be both to assess the degree of compliance with policies, requirement documents, and procedures and to assess their adequacy as well, with the following constraints:

- When a function, process or area is subject to frequent, periodic audit, it need not address the adequacy of the environmental programme. The large majority of the audits should address only the degree of implementation compliance with, and the adequacy of the programme. When the adequacy of the programme is to be addressed, there should be a higher-level overview by the management of the auditing organisation, i.e., an overview of any findings related to programmatic inadequacy. The purpose of this overview should be to assure that either each issue is fresh or each issue warrants a revisit before putting other units of the organisation through the perturbation of addressing the issues (Marguglio, 1991).

Note that a company's motivation for carrying out an environmental audit will determine the type of audit it chooses to implement. We will discuss the types of audits in Subsection 5.1.2. Before we do so, let us first do Learning Activity 5.1.

LEARNING ACTIVITY 5.1

Obtain a form for submission of Environmental Statement from local pollution control board office and list three benefits of this statement.

Note:

a) Write your answer in the space given below.

b) Check your answer with the one given at the end of this book.

5.1.2 Types of Environmental Audits

In this Subsection, we will discuss the two main types of environmental audits, i.e., objective-based and client-driven.

Objective-based Types

As mentioned earlier, environmental audit covers assessment of any activity that impinges on the environment. The scope and objectives of the audit more usefully distinguish different audit categories and how the audit results are to be used. However, you must note that objectives and scope are often a combination of several audit types and are usually defined on a case-by-case basis. Organisations have developed audit programmes to fit their particular needs. Based on objectives, environmental audits can be categorised as under:

(i) **Liabilities audit:** Compliance audit, operational risk audit, acquisition audit and health and safety audit form liabilities audit. These are often conducted as a prelude to gaining insurance cover and as a means of demonstrating the regulatory compliance. Compliance auditing is probably the most common form of environmental audits; it is a verification process whereby the facility establishes the extent to which it is complying with environmental legislations, regulations, emission limits, etc. Operational risk concentrates on the potential frequency and consequences of environmentally damaging activities in the raw material and product storage/handling and manufacturing process. Compliance with regulations does not necessarily reduce liability due to operational risks.

Acquisition audits assess the liability due to contaminated land and building remediation costs. Health and safety audits normally form part of health, safety and environment (HSE) audit and involve assessment of adequacy of personal protective equipment (e.g., safety shoes, goggles, helmets, etc.), emergency preparedness and disaster management plans.

(ii) **Management audit:** Corporate audit, system audit, policy audit and issues audit form management audit. These pay considerable attention to management systems as they guide the efficient and effective running of the operations. A corporate audit is initiated by the main Board of a parent company and is concerned with organisation structure, roles and responsibilities, policy implementation, awareness and communications with a subsidiary. This is carried out as a reassurance to the main Board that its aims and objectives are being implemented throughout the corporate structure. Management system audits are carried out to check the systems against the policy and standards such as British Standard 7750 or ISO 14001 (discussed in detail in Unit 7). Policy audit is carried out to review and reassess the relevance of policy in light of developments (legal, technical, financial) within the organisation and outside. Issues audit is carried out to establish environmental management plan and targets.

(iii) **Activities audit:** Site audit, waste audit, product audit and cross-boundary audit form activities audit. These cover auditing of select technical and management issues. Environmental site audit examines all aspects of the facilities performance with respect to the environment. It combines most of the elements of other types of EA and, when undertaken in depth, involve considerable time and cost. The waste audits are of two types. The first identifies and quantifies waste streams and is a precursor to waste minimisation programmes. The second type assesses waste management practices and procedures. Product audits cover several aspects of their environmental impacts through design, manufacture, use and disposal. Such audits are pre-requisites for identifying environmentally friendly products for 'Green Labelling' (explained in Unit 9). Cross boundary audits assess activities, which cut across departments or business units (e.g., transport and supply chain audits).

Figure 5.1 illustrates the objectives-based audit types.

Environmental Audit Types		
Liabilities Audits	**Management Audits**	**Activities Audits**
Compliance Audit	Corporate Audit	Site Audit
Operational Risk Audit	Systems Audit	Waste Audit
Acquisition Audit	Policy Audit	Product Audit
Health & Safety Audit	Issues Audit	Cross-boundary Audit

Figure 5.1 Environmental Audit Categories

Client-driven Types

The different types of audits are based on the client, who has commissioned or ordered the audit procedure:

(i) **Regulatory external audit:** This often entails an examination carried out by or for an environmental regulatory agency, with the goal of ensuring that a facility is meeting the relevant legislation and regulations. The regulatory agency can use the methodology of audit as a tool to systematically enhance its overview, including the possibility of verifying the accuracy of any reports, which a company is required to submit to the authority.

(ii) **Independent external audit:** This is conducted by external auditors entitled to perform audits. As the environmental factors have gained importance for a firm's market relations, shareholders such as banks and investment funds, insurance companies, environmental groups, potential buyers, customers, local government and environmentally aware citizens are demanding independent external audits to assess how the firm deals with environmental issues.

(iii) **Internal environmental audit:** This often involves an inquiry commissioned by management. In practice, such audits are commonly ordered by senior management located at some distance, in both physical and operational senses, from the factory or site of environmental concern. In such cases, the environmental audits are internal in that the results will remain within the organisation. However, for the facility under investigation, the internal audit will have the same effect as an external audit. One reason why firms conduct internal environmental audits is to diminish their liability to pay fines, damages or clean-up costs as the result of breaking the law (e.g., releasing more emissions than permitted).

(iv) **Third party audits:** These represent the audits certifying organisations carry out to verify as to whether internal/external audits meet the standards set.

In Table 5.1, we illustrate the client-driven audit types.

Table 5.1 Environmental Audit Types: Client-driven

Category of Audit	*Ordered by*	*Desired Result*
Regulatory External	Regulatory Authority	Enhanced oversight
Independent External	Buyer, bank, customer, insurance firm, etc.	Objective information
Internal	Top management, Members of Board	Reduced risk
Third Party	Top management, Members of Board	Certified environmental protection system

5.2 GENERAL AUDIT METHODOLOGY

What the foregoing discussion suggests is that the focus of environmental audit differs, depending on the specific requirements of the clients. In other words, audit methods and tools are tailored to suit the purpose for which they are intended. However, any audit programme must conform to the basic framework. We will explain the basic structure of, and the steps involved in, an audit programme in Subsections 5.2.1 and 5.2.2, respectively.

5.2.1 Basic Structure of EA

Through different production facilities and activities, audit orders and types of audits place different requirements on the structure of an audit, there are certain elements, which should be present in any effective environmental audit programme. These common elements include:

(i) Explicit top management support for environmental audit and commitment to follow-up on audit findings.
(ii) An environmental audit function independent of audited activities.
(iii) Adequate team staffing and auditor training.
(iv) Explicit audit programme objectives, scope, resources and frequency.
(v) A process which collects, analyses, interprets and documents information sufficient to achieve audit objectives.

(vi) A process which includes specific procedures to promptly prepare candid, clear and appropriate written reports on audit findings, corrective actions and schedules for implementation.

(vii) A process which includes quality assurance procedures to assure the accuracy and thoroughness of environmental audits.

5.2.2 EA Steps

The four general steps involved in an audit procedure are:

(i) **Audit preparation:** This includes choice of auditor/audit group, collection of background material and planning of the audit orientation. Audit preparation is crucial in determining the methodology, practical tools and materials that will be required for the completion of the audit. In many cases, the available materials such as manuals, questionnaires, etc. may not be well suited to the type of activity that is to be audited, or the type of audit that is performed. This means that part of the preparatory activities may have to involve development of audit-specific tools or further development of existing materials and tools. In such cases, this may require a preliminary review of the facility or activities to map out specific details and requirements to be observed in the development of the particular tools.

(ii) **A systematic scrutiny or review of a facility:** Depending on the orientation and goals of the audit, the focus of the scrutiny differs. For example, an environmental audit can be conducted without dealing with the processes involved in production. However, an audit, which is considered part of a programme of preventive environmental protection, needs to deal with the production processes and material flows. Common to all audits is an analysis and evaluation of the information that has been obtained and an analysis of the outcome vis-a-vis the goals and expectations of the initiator of the audit. This helps in identifying the areas of improvement.

(iii) **Reporting:** This step involves reporting of observation of deficiencies and possible alternatives. It is important to be aware that an environmental audit by itself does not solve any problems. In fact, during the work on the audit, it may appear that the environmental problem is increasing because the audit process brings to attention the hitherto unknown problems or deficiencies. Audits often point to the need for the changes in organisation and improvements in education, increased environmental responsibility, and investments in new equipment and environmental protection technology. An important pre-condition for the success of an audit, especially if it is internally initiated, is that everyone is prepared to accept the consequences and take steps to solve the deficiencies and problems, which the audit may reveal. Thus, the persons whose areas of responsibility have been the object of scrutiny must make decisions and have plans in place to eliminate the problems.

(iv) **Follow-up:** Following up of the results is an important part of the audit process. An evaluation of the results of the remedial actions is a logical step. This can be done either as part of the subsequent audit or as part of a continuing process of enhancing environmental protection procedures. However, we must note that mere following up on the results may result in a control routine and a stagnation in work on environmental problems. It is, therefore, important that a follow up process also analyses the methodology and orientation of the previous audits to ensure that it remains a dynamic factor.

Figure 5.2 presents the basic steps in an audit procedure.

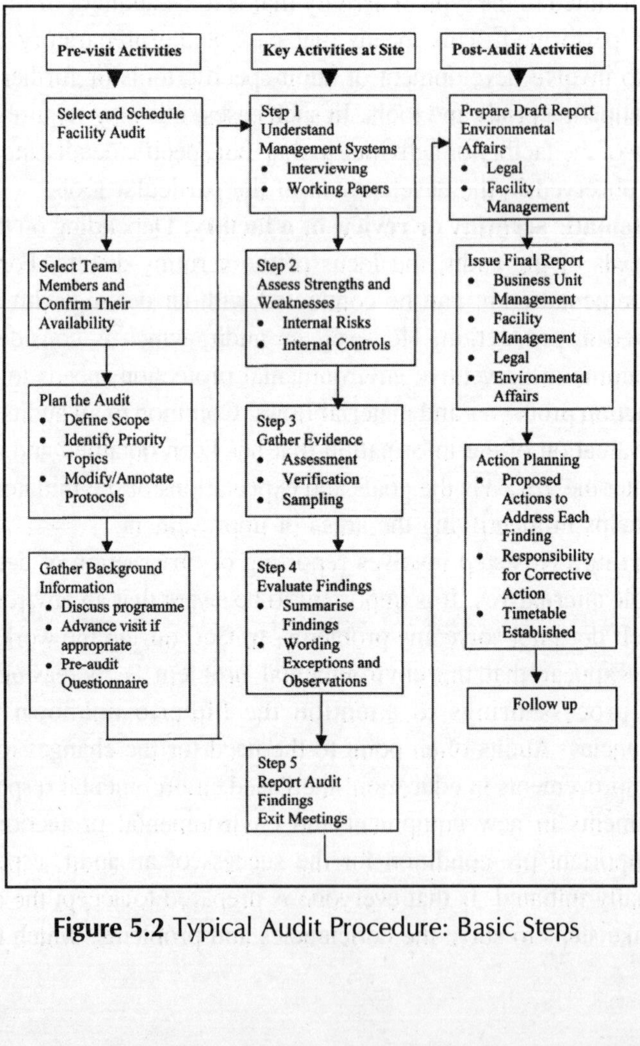

Figure 5.2 Typical Audit Procedure: Basic Steps

Having discussed the basics of an environmental audit, let us now look into the issues involved in planning an audit.

5.3 ELEMENTS OF AUDIT PROCESS

In this Section, we will discuss how to plan an environmental audit. In planning an environmental audit, it is important to decide about the scope of the audit within an organisation (given that an audit need not necessarily cover the entire organisation), audit team, objective of audit and audit methodology. We will discuss these in Subsections 5.3.1 to 5.3.4.

5.3.1 What to Audit?

An audit can cover anything from environmentally-related decisions to the function and maintenance of an individual piece of environmental protection equipment. In other words, concerning what should be audited, many definitions permit a rather broad approach to organisational, administrative and technical matters.

A typical result of preliminary audits is that some sort of organisational structure is created within the firm, or just at the production site and that some procedures for dealing with environmental problems are established. An obvious result of these actions is that future audits ought to evaluate whether the organisation and the procedures function in the ways that they were intended. A natural turn of these developments is that environmental audits become more and more management systems-oriented as a firm's environmentally-related activities become increasingly formalised and systematised.

For instance, in the management context, the transition from problem solving to quality assurance mechanisms, which attempt to prevent problems, is a natural development. The process seems to be no different in the environmental management context as well. The pattern of transition, in fact, is reminiscent of financial audit where a distinction can be made between records investigation and internal control evaluation. As methods of internal control are developed, for example, audits can concentrate increasingly on verifying those organisations, practices, procedures, the divisions of work and responsibility and other features which function as they were intended. Increasingly, in the same way, quality environmental audit is being equated with the audit of quality management systems, even where there are other types of quality audits, which are more technically-oriented towards processes and products.

5.3.2 Who Should Audit?

The question of who should conduct an audit remains open. A clear strategy on the part of industry has been to try to reserve the use of this tool for industry itself. On the other hand, regulatory agencies have seen the possibilities of audit in their efforts to improve their insight into the environmental impacts of production. These ambitions need not be in opposition with each other, rather, they should serve as complements. The questionnaires and checklists, which companies have developed prior to conducting an internal audit often show similarities to those that the regulatory agencies use during their periodic inspections and general evaluations.

In addition to environment regulatory agencies and firms themselves, there are many other actors in society who have an interest in how the environmental issues are dealt with in companies. These actors can use the tools and methodology of environmental audits to seek objective information. Increasing environmental consciousness among consumers, efforts to improve quality and development towards greater standardisation are the three factors that have led to the inclusion of environmental relations in quality parameters. In the future, it may become commonplace for various interested parties to examine not only production and product quality, but also environmental quality, to estimate the viability of the firm in the marketplace. Environmental audit may, therefore, be used by many parties to secure their interests. Thus, it should not be defined in a manner that limits its use to a particular group.

5.3.3 Why an Audit?

The definition of the primary purpose for conducting an environmental audit is often a reflection of priorities and interests. Industries, on the one hand, see environmental audits as means of reducing their own environmental risk taking. On the other hand, environmental groups and organisations emphasise that it is the risks to the environment that are to be minimised through environmental audit. A reduced risk of legal action against a firm, as the result of excessive emissions or accidents, may also spare the environment from certain types of pollution and waste.

There are also differences inherent in these perspectives about the purpose of environmental audits. For example, potential risks may be identified based on fulfilling legal requirements or from a purely environmental perspective. To consider everything that is not specifically forbidden by law as permissible could be a defensible position, which some firms do, provided that environmental regulations expressed some sort of optimal level of pollution. Yet, it is becoming more commonly understood that this is not the case. Pollution and waste always represent a loss of materials and an environmental burden, which also always entails costs for the firm and the environment.

A certain broad agreement on continued efforts and goals for environmental protection is developing. Concerned regulatory authorities, political parties, businessmen, unions, researchers, environmental organisations and increasingly environmentally aware consumers are tending to agree on the idea that sustainable development must be based on a strategy of preventive environmental protection. This strategy can be expressed in many ways, including the definition of the purposes of environmental audits.

5.3.4 How to Audit?

Finally, when it comes to conducting an environmental audit, it is natural to relate this to the methodology and practices used within financial and quality audits. There is no advantage to have environmental audits depart radically from established procedures in related areas.

Common to both financial and quality audits is the demand to be systematic, objective and independent. This means that competent individuals who do not have direct responsibility for the areas that are to be investigated conduct the audit. Yet, to ensure that the audit process is not associated with direct methods of control and supervision, the audit should be conducted with co-operation from the personnel who work in the area under investigation. An additional advantage of involving personnel in the audit procedure is that this may lead to a better understanding of why improvements and corrective actions are necessary.

Documentation and periodicity, i.e., the frequency of reports or analyses, is a general requirement in definitions of environmental audits, even if they may seem to be redundant. The need for documentation of the audit procedure and its results should be obvious, while periodicity is a function of the purpose of the audit. If the primary purpose, for example, is to ensure that a dynamic process continues in the direction of diminishing environmental **bidets**, then a high frequency of audits may be required.

LEARNING ACTIVITY 5.2

Say whether it is important that environmental audit should be an ongoing process.

Note:

a) Write your answer in the space given below.

b) Check your answer with the one given at the end of this book.

____

____

____

____

____

____

____

5.4 WASTE AUDITS AND POLLUTION PREVENTION ASSESSMENTS

As mentioned earlier, waste audits are intended to identify the possibilities for improvement. The recommendations contained in waste audits are not based on objective comparisons between expectation and results, but instead are based on technical and economic evaluation of various waste minimisation alternatives.

The origin of the term *waste audit* can be traced to the amendment of the Resource Conservation and Recovery Act (RCRA) enacted in the USA during the early 1980s. The RCRA was revised because of the problems involved in the handling and land filling of hazardous wastes. The work on revising the RCRA resulted in the Hazardous and Solid Waste Amendment of 1984, in which the US Congress declared the highest priority to be preventive measures.

The amendments required all large producers of hazardous wastes to put a programme in place to reduce the volume and toxicity of wastes. Further, waste producers were required by law to submit biennial reports to the environmental authorities on the results of efforts to reduce waste volume and toxicity.

This legislation led to a widespread introduction of waste minimisation programmes within industries to identify the potential for reducing wastes at source. These minimisation programmes required systematic procedures to localise and characterise all types of waste streams, and this led to the creation of formal methodologies for waste minimisation audits. The US Congress Office of Technology Assessment (OTA) defines waste audit as a systematic, periodic internal review of a company's processes and operations, designed to identify and provide information about opportunities to reduce wastes (OTA, 1986).

This last aspect of audits raises the question of the auditor's freedom to provide advice or recommendations to the organisation that is the object of scrutiny. An advisory capacity is often thought to be in conflict with the requirement of objectivity, because the auditor runs the risk of examining his or her own proposals in a future audit. Yet, this has not hindered the making of recommendations by those involved with financial or quality audits. The audit process assumes that the competent and experienced auditor recommends improvements in an objective and independent way. This means that it must be clear to all parties involved in an audit that the process of environmental audit does not assume any part of the operative responsibility for environmental protection that is the prerogative of management.

Note that waste audit encompasses the term *waste* but does not refer to merely hazardous and other solid waste; it refers to all by-products of production, including emissions to the air and water. UNEP and UNIDO are promoting waste audits as an operational management tool. Figure 5.3 shows a schematic presentation of their recommended work procedure.

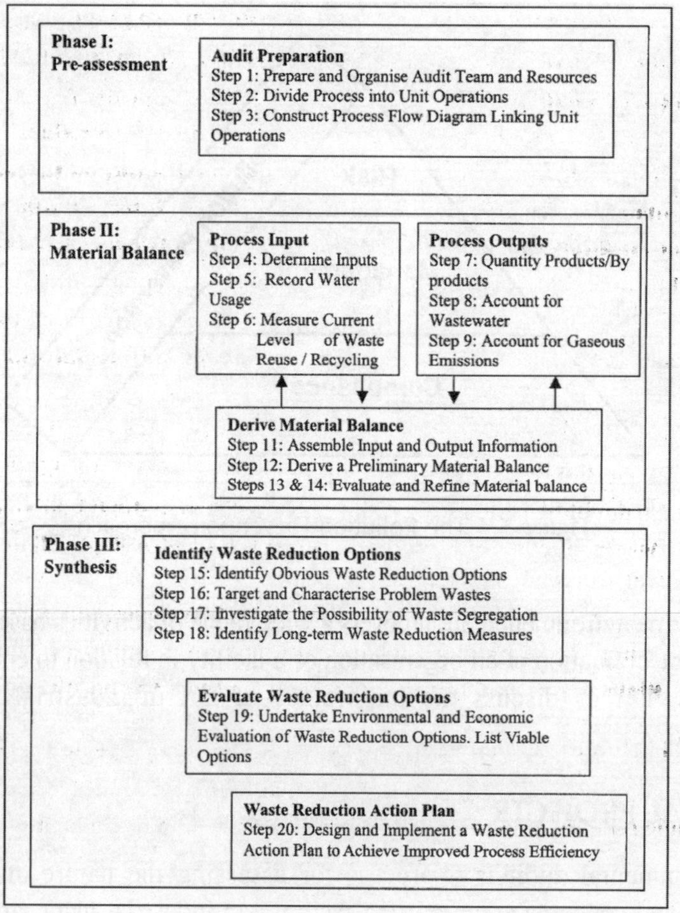

Figure 5.3 Waste Audit: A Guide

Audits, which contain an explicit pollution prevention approach, are considered to be pollution prevention audits. This classification is used regardless of whether the audit is conducted independently, or the pollution prevention approach is integrated into compliance, management, and/or risk audits, as represented in Fig. 5.4.

Note that pollution prevention audits are operative in the sense that they include the creation and analysis of correction and improvement alternatives. In formal environmental audit, this is a task reserved for management. Furthermore, pollution prevention audits and similar audits require the active involvement of management in information gathering and proposals for change. For reasons such as these, in this context, the term *assessment* is considered to be more appropriate than audit.

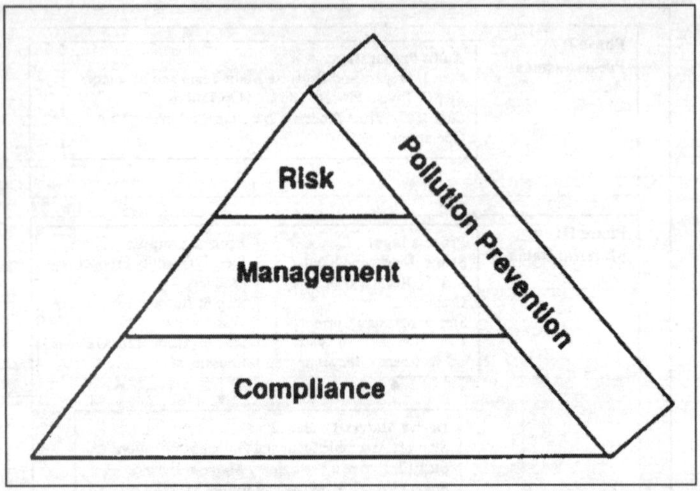

Figure 5.4 The Pollution Prevention Approach

The term environmental audit covers a wide range of activities based on formal performance evaluation of an organisation or a facility in relation to environmental objectives. Let us discuss the importance of EA in industrial projects in Section 5.5.

5.5 EA IN INDUSTRIAL PROJECTS

An environmental audit is a process for assessing the nature and extent of environmental concerns at an existing facility—an industrial plant, an abandoned site, a mine area or any other site, where industrial pollution problems are identified or anticipated (Pollution Prevention and Abatement Handbook, 1998). It is used to provide data on the extent of pollution in an industrial area, to quantify the scale of pollution at a particular site or to examine the causes and potential remedies of problems at a facility. This Section provides guidance on the uses of environmental audits in industrial pollution management and on the scope of a typical audit.

Site Audit

A *site audit* is often the first step in obtaining a quantitative understanding of pollution problems. In many cases, the audit allows an evaluation of priorities, the extent and cost of control and remediation measures. This information then shapes all remediation actions and investments.

In an industrial context, the overall objective of a site audit is to understand the scale and sources of the pollution problems at a facility or in a defined area and to set out the options available for dealing with those problems. This is often a staged process of investigation in which each stage is narrower in scope but more detailed than the preceding one. An initial assessment can be relatively quick, drawing on readily available sources, including site interviews, and providing an overview of the actual or suspected sources of pollutants and the extent of their impact. This overview can be carried out during project definition or in the *scoping* stage and provides a basis for further detailed investigations or for defining priorities for action. An initial assessment also helps describe or indicate where site sampling and monitoring might be cost-effective.

A *full site audit* is detailed, requiring careful site inspections, (perhaps, including sampling and testing) and review of past and present production processes, as well as pollution emissions and control measures. The audit should also clarify the legal and regulatory framework, licensing agreements, corporate policies, and management structures and priorities that affect the environmental performance of the plant. In many cases, relevant technical and environmental standards for performance may be ill defined or may not exist, and professional judgements will have to be made as the appropriate benchmarks. However, it is essential that the standards or emissions limits proposed for the plant be clearly defined and that the rationale for their selection be given. If plant requirements appear unachievable with the current plant, the audit should address what might be acceptable as realistic interim requirements.

Audit Purpose

An audit should provide a list of recommended actions, in terms of increasing cost-effectiveness in addressing the critical environmental issues. This list should include interim and long-term targets and a timetable for achieving them, together with indication investments and other resources (e.g., humans, information, and so on) that would be required.

The scale and cost of site audits can vary widely, depending, above all, on the extent of field data collection required. One or two persons can carry out a scoping study in a couple of days, with cooperation from site personnel. A full site audit of a typical industrial plant can be carried out within one to two months, of which perhaps one week is spent on the main fieldwork, usually followed by a shorter visit to verify initial findings.

Terms of Reference

The terms of reference (TOR) of a site audit will obviously be site-specific. However, a typical TOR includes the following:

- Overall objective.
- Specific objectives.
- General scope of work.
- Baseline data.
- Principal sources of contamination.
- Area of impact.
- Technical approach.
- Work plan.
- Prevention, control, and mitigation.
- Recommended priority actions.
- Environmental management plan.
- Site-specific scope of work (by site).
- Laboratory services.
- Norms and standards.
- Pre-bid site visits.
- Client contacts and counterparts.
- Facilities provided by the client.
- Reporting and deliverables.

A site audit report may contain the following sections:

- Executive summary.
- Project description.
- Regulatory setting.
- Audit procedure.
- Mitigation.
- Costs and schedule.
- Annexure.

Points of Procedure

The following points relate to the procedures for the execution of a site audit.

(i) **Selection of auditors:** Various forms of certification of environmental auditors are under discussion in different countries. Although many of the skills required for a site assessment are general environmental or engineering skills, it is important that the audit team has personnel with detailed knowledge of the specific industry being addressed. The selection of auditors should follow the normal procedures for consultants, and arrangements should be made to allow bidders to become familiar with the site before the tender closure date.

(ii) **Briefing and TOR:** It is essential that the consultants selected have a clear understanding of the objectives of the work, especially if the audit is to become part of the overall environmental assessment for the project. The TOR, therefore, needs to be as specific as possible.

(iii) **Preparation phase:** An audit plan should contain a description of the information required, the site visit schedule and the site personnel to be involved or interviewed. A protocol may also be prepared defining the specific information that will be sought during the site visit. The protocol should be provided to the enterprise well in advance of the visit. Available file information on the facility should be obtained and reviewed before the visit, and the audit plan should then be refined, if necessary.

(iv) **Execution of the audit:** Active cooperation of the plant owners and managers is essential for a good result and should be secured in advance. As effective coordination reduces delays and costs, site visits, interviews, and any sampling should be organised as early as possible. The site inspection should be carefully documented to support the findings and recommendations and to provide a reference for future audits.

(v) **Review of findings:** It is important that the management in place be allowed to comment on the findings and recommendations of the audit.

Our discussion so far may give us an impression that environmental audits are relevant only to industries. In reality, however, these are useful in other sectors as well. For example, waste audits are used to identify a potential site contamination. Let us discuss this aspect in Section 5.6.

LEARNING ACTIVITY 5.3

State the purposes of a waste audit.

Note:

a) Write your answer in the space given below.

b) Check your answer with the one given at the end of this book.

5.6 LIABILITY AUDITS AND SITE ASSESSMENT

With the advent of stringent legislation on liabilities for contaminated soil and groundwater, there has been a growing demand for information on the environmental state of properties and the potential for on- and off-site pollution migration.

In the United States, for example, 100,000 sites have been labelled as contaminated, of which 10,000 have been described as priority cases, and these properties are called *brownfields*. The US Environmental Protection Agency (EPA) defines brownfields as abandoned, or under-used industrial and commercial facilities where expansion or redevelopment is complicated by real or perceived environmental contamination. Contaminated land is also fairly common in Central and Eastern Europe and in industrialised regions in many developing countries like India.

Many of these sites offer attractive financial opportunities, but a major obstacle to their redevelopment is the fear of contamination (left by past users) and associated liabilities. Major liabilities include potential remediation/cleanup; regulatory compliance; fines and penalties for regulatory non-compliance and compensation to private parties. Under the US Comprehensive Environmental Response, Compensation and Liability Act (CERCLA), often referred to as Superfund, the current owner of a contaminated site may be held liable for the cost of clean up. In India, the responsibility of the occupier and operator of a facility handling hazardous wastes is also defined under hazardous waste (Management & Handling Amendment Rules, 2000).

Often, the costs of investigations and remediation measures exceed the value of the property. Investigations to identify actual or potential site contamination are called site due diligence or liability audits. At times they are referred to as environmental site assessments (ESA).

5.6.1 Liability Audit: Phase 1

The term *Phase 1 liability audit* is used for investigations that involve collecting information from interviews, by studying available historical information and by performing visual inspections of sites. It is also known as *pre-acquisition site assessment*, *preliminary site assessment*, *liability audit* and *environmental due diligence*. In general, Phase 1 audit is carried out based on the country specific guidelines or standards, as illustrated below:

- American Society for Testing and Materials (ASTM) Standard E 1527-97 entitled *Standard Practice for Environmental Site Assessments: Phase I Environmental Site Assessment Process.*

- ASTM Standard E1528-96 entitled *Environmental Site Assessments: Transaction Screen Process*.
- The Interdepartmental Committee (of the UK Department of Environment) on the Redevelopment of Contaminated Land (ICRCL) guidance note 59/83 (2nd Edition, July 1987).

In practice, it is indeed often necessary to take a phased approach, with a Phase 2 audit following a Phase 1 audit. Phase 1 audits, which are relatively inexpensive and quick, can help screen out those sites that do not require further investigation, thereby reducing the uncertainty of potential environmental liabilities.

The audits should also identify the needs for, and possibly the scope of, more thorough investigations in the form of a Phase 2 audit. The findings in the Phase 1 audit should state as to whether or not contamination is suspected as well as the likelihood of contamination. This approach can limit the need for drilling, sampling and analysis, and thereby reducing costs. A Phase 1 audit should also place the site-specific findings into the context of the surrounding environment.

5.6.2 Liability Audit: Phase 2

Phase 2 liability audit refers to investigations that involve detailed physical sampling and testing of contaminants in laboratories.

The outcome of the Phase 1 audit forms the basis of the exploratory subsurface investigation (Phase 2). This involves drilling, sampling and analysis of primarily soil and groundwater and assessment of respective contamination against clean up standard (e.g., The Dutch Ministry of Public Housing, Land Use and Environmental Guidelines Soil and Groundwater Standards). This also involves collection of geological and hydro-geological data of the site to establish lateral and vertical extension of the contaminants, if any. Note that in the absence of any soil and groundwater standards, this or other country-specific standards can be adopted in India for assessment purpose.

Phase 2 audit helps determine the need for additional investigations or remediation. If contamination is identified through this process, the question of remediation and clean up is dictated by a number of factors such as legislation, future land use, risks of contaminant spreading and possible impacts on human health and the environment. If necessary, a remediation plan is developed as part of, or as follow-up to, the Phase 2 audit.

The remediation activities are sometimes termed as Phase 3 of the process. It is important to strike a balance between the time and costs of investigations and the need for additional information to reduce uncertainties about possible contamination. The uncertainties inherent in site investigations make it crucial to document observations, findings and how conclusions have been formed. It is also important to use the information obtained in one part of an audit (e.g., the site inspection) to confirm or invalidate observations from other parts (e.g., record reviews or interview sessions).

Typically, the scope of liability audit includes the following:

- Initial briefing sessions at the company's facilities between the consultant and appropriate company staff.
- Review of relevant environmental and occupational health and safety legislation.
- Review of existing documentation pertinent to all of the facilities and the environmental and occupational health and safety aspects of the facilities.
- Inspection of all buildings and properties in which significant manufacturing, laboratory, or chemical storage/disposal operations are to be included in the audit.
- Identification of all environmental and occupational health and safety concerns related to both past and ongoing activities.
- Preparation of a prioritised list (i.e., high, medium and low) of concerns related to past and ongoing activities.
- Providing recommendations and estimated costs on what additional remediation measures are required for both past and ongoing environmental concerns.
- Preparation of a report that identifies all relevant environmental and occupational health and safety legislations/concerns; prioritisation of all concerns related to past activities as also ongoing activities; recommendations on what further action is required along with a cost estimate for such actions for both past and ongoing activities and an executive summary highlighting the key findings.

LEARNING ACTIVITY 5.4

State the type of EA you would like to conduct in your area and list the required steps for implementing EA.

Note:

a) Write your answer in the space given below.

b) Check your answer with the one given at the end of this book.

5.7 AUDITING OF EMS

An environmental management system (EMS) audit refers to a systematic and documented verification process of objectively obtaining audit evidence to determine whether an organisation's EMS conforms to the audit criteria and communicating the results of this process to the client *(http://www.pdfwebsearch.com/regscan/unprotected/ISO+14011-1996.pdf)*.

The EMS audit criteria includes policies, practices, procedures or requirements, such as those covered by ISO 14001 and, if applicable, any additional EMS requirements against which the auditor compares collected audit evidence about the organisation's environmental management system.

5.7.1 Audit Objectives, Roles and Responsibilities

The objectives of a typical EMS audit include:

- determining conformance of an auditee's EMS with the EMS audit criteria;
- determining whether the auditee's EMS has been properly implemented and maintained;
- identifying areas of potential improvement in the auditee's EMS;
- assessing the ability of the internal management review process to ensure the continuing suitability and effectiveness of the EMS;
- evaluating the EMS of an organisation where there is a desire to establish a contractual relationship (e.g., with a potential supplier or a joint-venture partner).

Auditors and clients have an important role to play in auditing environmental management systems.

5.7.2 Scoping and Review

The scope of the audit describes the extent and boundaries of the audit in terms of factors such as physical location and organisational activities as well as the manner of reporting. The client and the lead auditor determine the scope of the audit. The auditee should normally be consulted when determining the scope of the audit. Any subsequent changes to the audit scope require the agreement of the client and the lead auditor. The resources committed to the audit should be sufficient to meet its intended scope.

At the beginning of the audit process, the lead auditor should review the organisation's documentation such as environmental policy statements, programmes, records or manuals for meeting its EMS requirements. In doing so, appropriate background information on the auditee's organisation must be used. If the documentation is judged to be inadequate to conduct the audit, the client should be informed. Additional resources should not be expended until further instructions have been received from the client.

5.7.3 Preparing the Audit

The audit plan should be designed to be flexible in order to permit changes in emphasis based on information gathered during the audit and to permit effective use of resources. The plan should include most of the following elements:

- the audit objectives and scope;
- the audit criteria;
- identification of the auditee's organisational and functional units to be audited;
- identification of the functions and/or individuals within the auditee's organisation having significant direct responsibilities regarding the auditee's EMS;
- identification of those elements of the auditee's EMS that are of high audit priority;
- the procedures for auditing the auditee's EMS elements as appropriate for the auditee's organisation;
- the working and reporting languages of the audit;
- identification of reference documents;
- the expected time and duration for major audit activities;
- the dates and places where the audit is to be conducted;
- identification of audit team members;
- the schedule of meetings to be held with the auditee's management;
- confidential requirements;
- report content and format, expected date of issue and distribution of the audit report;
- document retention requirements.

The audit plan should be communicated to the client, the audit-team members and the auditee. The client should review and approve the plan. If the auditee objects to any provisions in the audit plan, such objections should be made known to the lead auditor and resolved between the lead auditor, the auditee and the client before conducting the audit. Any revised audit plan should be agreed between the parties concerned before or during the audit.

5.7.4 Conducting the Audit

Audits are conducted with an opening meeting to:

- introduce the members of the audit team to the auditee's management;
- review the scope, objectives and audit plan and agree on the audit timetable;

- provide a short summary of the methods and procedures to be used to conduct the audit;
- establish the official communication links between the audit team and the auditee;
- confirm that the resources and facilities needed by the audit team are available;
- confirm the time and date of the closing meeting;
- promote the active participation by the auditee;
- review relevant site safety and emergency procedures for the audit team.

Sufficient audit evidence should be collected to be able to determine whether the auditee's EMS conforms to the EMS audit criteria. Audit evidence should be collected through interviews, examination of documents and observation of activities and conditions. Indications of non-conformity to the EMS audit criteria should be recorded. Information gathered through interviews should be verified by acquiring supporting information from independent sources, such as observations, records and results of existing measurements. Non-verifiable statements should be identified as such. The audit team should examine the basis of relevant sampling programmes and the procedures for ensuring effective quality control of sampling and measurement processes, used by the auditee as part of its EMS activities.

After completion of the audit evidence collection phase and prior to preparing an audit report, the audit team should hold a meeting with the auditee's management and those responsible for the functions audited. The main purpose of this meeting is to present audit findings to the auditee in such a manner as to obtain their clear understanding and acknowledgement of the factual basis of the audit findings. Disagreements should be resolved, if possible, before the lead auditor issues the report. Final decisions on the significance and description of the audit findings ultimately rest with the lead auditor, though the auditee or client may still disagree with these findings.

5.7.5 Content of Audit Report

The audit report should be dated and signed by the lead auditor. It should contain the audit findings and/or a summary thereof with reference to supporting evidence. Subject to agreement between the lead auditor and the client, the audit report may also include the following:

- the identification of the organisation audited and of the client;
- the agreed objectives, scope and plan of the audit;
- the agreed criteria, including a list of reference documents against which the audit was conducted;
- the period covered by the audit and the date(s) the audit was conducted;
- the identification of the auditee's representatives participating in the audit;
- the identification of the audit-team members;

- a statement of the confidential nature of the contents;
- the distribution list for the audit report;
- a summary of the audit process including any obstacles encountered;
- audit conclusions such as EMS conformance to the EMS audit criteria, whether the system is properly implemented and maintained and whether the internal management review process is able to ensure the continuing suitability and effectiveness of the EMS.

The audit is completed once all the activities defined in the audit plan have been concluded.

SUMMARY

The focus of this Unit was on the emergence of environmental audit as a proactive environmental management tool. We began the Unit by giving you an overview of the objectives and scope of an environmental audit. We then discussed the methodologies of and procedures involved in carrying out an audit programme. We also discussed the various applications of environmental audit and their scope and in that context explained waste, liability, industrial and EMS audits.

REFERENCES

ISO 14001 www.iso.org

Marguglio B. W. 1991. Environmental Management Systems, Marcel Dekker, ASQC Quality Press, New York, Milwaukee

OTA 1986, Opportunities for research in hazardous waste minimization, http://fas.org/*ota*

Prevention and Abatement Handbook, 1998: Toward Cleaner Production United Nations Environment Programme, World Health Organization, World Bank

SUGGESTED READINGS

British Standards Institute BS 7750:1994. *Specification for Environmental Management Systems*.

Edward, J.R. (undated). Taking Environmental Audit to the next level: Moving to Pollution Prevention Audits in the 1990s. U.S.EPA.

International Chamber of Commerce. 1991. ICC Guide to Effective environmental audit, Paris.

U.S. Environmental Protection Agency. 1986. Waste Minimization Opportunity Assessment Manual, Cincinnati.

CASE 5.1: WASTE AUDIT: THE VISCOSE RAYON UNIT CASE

Preamble

Viscose Rayon in Mumbai manufactures 270 TPD of Rayon Grade pulp, Viscose staple fibre, and Viscose filament yarn. Material, energy, health and safety audits were conducted to identify avenues for savings in the cost of production. Measures suggested herein, if implemented, are envisaged in savings to the tune of Rs 125 lakhs per annum.

Recommendation Based on Material Audit

In Chipping unit, the undersized chip pieces and the sawdust during chipping process amount to about 10 kg/1000 kg of wood charged. But due to the operational problems and design of the control system, the waste material is not recovered through the blower system for reuse in the boiler as fuel. Some of the material is lost as waste solid. By implementing a proper dust collecting system, the chipped material can be recovered in larger quantities to reduce the solid waste in the chipper section. At present, the loss of wood in the chipper section is about 1 to 2%. This can be reduced to about 0.5 to 0.75% through proper waste collection and reuse systems.

In the pulp plant, about 150 tons of pulp is produced per day. The pulp plant consists of chipper, digester, extraction and drying units. During liquor washing, loss of pulp is identified as 1%, which can be reduced to 0.6–0.7% through good housekeeping. The major loss of pulp is in the screening section, which is about 5% due to oversize of cooking material. The washed oversized material can be collected and reused.

Four TPD of pulp is wasted during bleaching process as rejects in the pulp plant. The same can be collected and processed further for second grade pulp. This is economically viable.

In digesters, SO_2 concentration of CBS should be maintained around 5.0–5.5% for reducing the digestion time and increasing the number of batches per day, which is economically beneficial.

In the pulp plant, an excess amount of about 0.04 T of sulphur/T of pulp can be saved if stochiometric quantities of chemicals are used for CBS preparation.

Recommendations Based on Energy Audit

The plant consumes electrical and thermal energy for the production of staple fibre and rayon yarn. The annual energy consumption of the plant is about 111.8 MkWh of Electricity, 16,659 KL of furnace oil, 19,016 KL of LSHS, and 1498

KL of diesel. The plant invests Rs 9.91 crores worth of electrical energy apart from self-generation, and Rs 22.6 crores worth of thermal energy.

The audit study has identified energy saving opportunities to an extent of 9.8 MkWh of electrical energy, and 542 KL of furnace oil per year. The total estimated savings in energy bill is Rs 205.87 lakhs per year, which is 6.33% of the purchased energy cost.

A detailed analysis of the electrical system has identified following areas for energy conservation:

(i) Improving power factor to 0.92 by providing additional capacitors at the main 110 KV outdoor station, and operating the DG sets at optimum power factor.

(ii) Achieving the optimum loading on the effluent treatment plant and power plant substation transformers.

(iii) The evaluation of performance and loading of various types of electric drives in the plant show energy conservation opportunities as follows:

  (i) Replacement of underloaded motors in agitators, exhaust blowers, pumps, driers and air supply fans.

  (ii) Providing electronic devices for energy savings to control power factor in slurry prcsscs.

  (iii) The lighting system in the Rayon Spinning, Rayon Spindle Expansion, Textile Spinning, and pulp plant shows following areas for energy savings:

  (iv) Switching off tube lights during 8 AM to 5 PM in Rayon spinning and Rayon Spindle Expansion plant.

  (v) Providing individual switches at quality inspection tables.

  (vi) Providing CFLs in place of bulbs and tube lights.

  (vii) Boiler test was conducted and the efficiency was found to be around 85%. The energy saving opportunity exists in sensible heat recovery from boiler blowdown.

The following major areas are identified for energy conservation in the process unit:

(i) Heat recovery from hot ammonia refrigerant gas in CS_2 plant.

(ii) Controlling CS_2 plant cooling tower fan by temperature sensor.

(iii) Minimising radiation and convection losses in VSF driers 1, 2 and 3, and rayon tunnel drier by increasing insulation thickness.

(iv) Recovering sensible heat by returning condensate from spin bath concentrator and tunnel drier to the boiler.

(v) Raising steam pressure in H_2SO_4 plant, waste heat boiler, and supplying HP steam to fibre driers.

(vi) Recovering sensible heat from FFFF evaporator foul condensate.

The following areas were identified for energy conservation in compressed air system:

(i) Cleaning of intercoolers.

(ii) Recovering heat from after cooler.

Recommendations Based on Water Audit

Total water requirement is 624 m^3/day in chipper house. The segregation of ejector and pump cooling water from digester house (264 m^3/d), and reusing the same in chipper house flume, reduces the total requirement by 360 m^3/day which saves Rs 96,000 per annum, whereas installation of piping for reuse costs around Rs 20,000.

Total freshwater requirement in first screen area is 6000 m^3/d excluding the warm water requirement. The backwater overflow from intermediate pulp storage chest is contaminated only with fibres, and hence 3100 m^3/day can be recycled for pulp dilution in common screen.

Recommendations Based on Environmental Quality Audit

Stack emissions

Gaseous pollutants like SO_2 and NO_x from pulp plant boiler stack exceed the CPCB standards. Higher SO_2 content is due to higher percentage of sulphur content in the liquor which is used as a fuel in boiler, whereas NO_x and CO show the incomplete combustion of fuel. For improved combustion techniques following measures should be adhered to:

(i) Proper burner maintenance.

(ii) Good atomisation of liquid fuel.

(iii) Provide bio-scrubbers or wet-scrubbers for controlling SO_2 content in the emissions.

Ambient air quality

(i) The SPM, SO_2 and NO_x concentrations were below the limits prescribed by CPCB for industrial mixed zones. Higher concentrations were observed in downwind directions.

(ii) Work zone monitoring results show that at all places concentrations are within the limits prescribed by TNPCB. However, further reductions are possible through control of leakages in the process. It is suggested that leakages in the process are detected and measures taken to control the same to ensure conservation of raw materials.

(iii) During the audit, odour problem was discernible. It is thus advised that SIV takes odour control measures through scrubbers.

Wastewater

(i) During study period, two equalisation basins were in operation. Running of two equalisation tanks concurrently is not required.

(ii) As a standard practice, airflow rate in equalisation basin should be in the range of 0.01 to 0.0115 m^3 of air/m^3 of tank to maintain the solids in suspension. So, for a tank of 9500 m^3, the airflow rate should be 5700 m^3/hr. Considering the capacity of existing air blowers, i.e., 2450 m^3/hr each, even if two blowers are operated simultaneously, the air supply will be insufficient to keep solids in suspension, resulting in accumulation of solids in equalisation basin.

(iii) Wastewater from equalisation basin is pumped to clariflocculator by three pumps (one standby). Capacity of each pump is 1500 m^3/hr. For optimal performance, it is required to pump wastewater at a constant (design) flow rate for which existing configuration of pump is not appropriate.

(iv) At present, lime is used to raise pH to 7 at equalisation basin for subsequent precipitation of zinc in clariflocculator. But, for optimum zinc removal, pH should be between 9 and 9.5. Therefore, due to low pH conditions, zinc removal efficiency is low. With the existing treatment scheme, it is technically not feasible to meet prescribed limit for zinc of 1 mg/L by lime precipitation alone.

(v) In case of aerated lagoons, the flow is not being divided properly and sufficient submersion for aerators is not kept. Hence, the desired effluent standards are not being met.

(vi) Provision of preclarifier is necessary for pulp plant effluent. Total effluent from pulp section is 1600 m^3/hr. Providing one hour detention time, a preclarifier of 20 m x 30 m x 3 m should be constructed wherein the fibrous material can settle. The cost savings are Rs 200/T. With this provision, frequent cleaning of equalisation tank will be avoided (presently cleaning is done two times in a year spending around Rs 1.50 lakh).

(vii) During the audit, it was observed that some of the weir plates in clariflocculators were missing which need to be replaced immediately.

(viii) As the existing effluent treatment plant cannot meet the stipulated standards, it is suggested that the extended aeration process is provided (with a settling tank after aerated lagoons I and II). A circular tank of 40 m diameter and 3 m depth can be constructed along with recirculating arrangement for the sludge to maintain 4000 mg/L of MLSS concentration. Excess sludge can be sent to the existing vacuum filter. With this, 95% BOD_5 can be removed and zinc concentration will be less than 1 mg/L in the final effluent.

(ix) In rayon plant, nest position in most of the machines are observed to be corroded, and nest water is leaking into the bottom of spinning machine, which must be rectified immediately.

(x) Section-wise norms are to be firmed for water consumption and strictly adhered to.

(xi) Aiming at techno-economic feasible solution, it is suggested that the following streams be segregated, and appropriate treatment given:

- High concentration zinc bearing effluent from VSF and Rayon sections.
- Acid condensate stream containing high BOD from wood pulp plant.

(xii) The viscose staple fibre and rayon plants discharge the effluent with rich content of zinc. Zinc can be precipitated from zinc bearing solution by the addition of lime and maintaining a pH of 9 in the treated solution or by using NaOH at a pH ranging from 9.3 to 9.5. Based on these principles, a method has been devised in which the raw wastewater is first treated with lime to raise its pH to 6, and the treated solution is settled to remove the mixed precipitates of calcium sulphate and zinc hydroxide. The supernatant is then treated with caustic soda and its pH adjusted to 9.3 when almost complete precipitation of zinc as hydroxide occurs.

(xiii) Colour removal from final effluent is a must before discharging to River Bhavani, warranting a physio-chemical treatment unit.

(xiv) As extended aeration must be practiced, there is no need of treating the effluent again in Aerated Lagoon III. It can be directly discharged into river Bhavani. By stopping all seven aerators in aerated lagoon III, a saving of Rs 43,45,000 per annum can be achieved.

Noise

(i) In high noise areas like boiler and H_2SO_4 plants, acoustic chambers with automatic spring doors should be constructed.

(ii) Near boiler, compressor room and new H_2SO_4 plant noise levels were observed to be higher than 90 dB. Workers in these areas should be mandated to use earmuffs.

(iii) Although the SIV has provided ear protective muffs, it was observed that most of the workers do not use them. High noise areas must be declared as Ear Protective Areas.

(iv) Noise-effect awareness programmes should be arranged among the workers.

Solid/hazardous wastes

(i) Sludge generated from the ETP at SIV falls under hazardous waste category No. 12.

(ii) Sludge has been characterized for leachate as well as its constituents. Yearly consumption of heavy metals (copper, zinc, lead, cadmium, nickel and chromium) exceed the quantities defined in Hazardous Waste Management and Handling rules, 1989, Govt. of India. The recovery of zinc from ETP sludge must be practiced to minimise expenses on secure landfill disposal.

(iii) Quantity of ash generated in boiler house due to combustion is 19 T/day. At present, it is being disposed without any precautions. The disposal of this sludge must be in a secure landfill.

(iv) Filter clothes from all sections in VSF are being dumped adjacent to the respective sections. These filter clothes contain viscose and caustic soda. During rains, these chemicals will reach groundwater deteriorating its quality. Thus the filter clothes should be disposed in a secure landfill.

Recommendations Based on Health and Safety Audit

(i) It is recommended that the safety function be reinforced to lay special emphasis on occupational health aspects such as in work areas toxins.

(ii) The storage of chlorine in the pulp plant needs a closer examination, and appropriate provision to deal with emergencies.

(iii) The safety organisation, though adequate to meet the requirements of Factories Act, does not possess expertise and facilities for analyses and investigations on work area toxins. It is recommended that a qualified person, with specific responsibility of dealing with the occupational health, be employed forthwith.

(iv) Accidents should be avoided while feeding logs to the flume in chipper section of pulp plant.

(v) A mild chlorine leakage in bleaching section of pulp plant was observed during the audit. Such incidents must be avoided.

(vi) Major (reportable) accidents are caused in wood yard and chipper section (50%) and sheeting section including baling section (35%). Hence, workers need proper training in material handling to avoid accidents.

(vii) During audit, broken door glass, broken gadgets, broken funnel, blocked spinneret, damaged pot cover in rayon spinning section, worn out flange packing in VP section, and leaking valves in spinning section were observed which are unsafe conditions, and must be rectified at the earliest.

(viii) Slippery floor in CS_2 plant and damaged step of a ladder in rayon bleaching section can be classified under poor house keeping conditions which require immediate attention.

(ix) Wet floor and spurting of hot pulp from alkali tower causes unsafe conditions in pulp plant.

(x) More stress on communication between the management and employees on safety is essential so that it creates an awareness leading to employees' acceptance of the safety programme.

UNIT 6

Life Cycle Assessment (LCA)

STRUCTURE

OVERVIEW

In Unit 5, we discussed, among others, how audits help improve the environmental performance of an industry. In Unit 6, we will discuss yet another tool for environmental management, i.e., life cycle assessment (LCA). We will first trace the evolution of LCA and describe the stages of a product life cycle. We will then identify the elements that contribute to the code of good conduct for an LCA. We will also explain the main steps involved in the LCA process, i.e., goal setting, inventory analysis, impact assessment and profile evaluation. We will close the Unit by explaining the application of LCA in private and public sectors.

LEARNING OBJECTIVES

After completing this Unit, you should be able to:

- explain the concept of life cycle assessment (LCA) as an environmental management tool and its potential for identifying all the environmental impacts throughout the entire life cycle of a product;
- describe what a code of good conduct for LCA entails;
- discuss the basic steps involved in an LCA process;
- conduct/co-ordinate an LCA and critically evaluate its outcomes.

6.1 EVOLUTION OF LIFE CYCLE ASSESSMENT

Life cycle assessment (LCA) is a tool to evaluate the environmental effects of a product or process throughout its entire life cycle. An LCA entails examining the product from the extraction of raw materials for the manufacturing process, through the production and use of the item, to its final disposal, and thus encompassing the entire *product system*. A schematic representation of a product life cycle is given in Fig. 6.1.

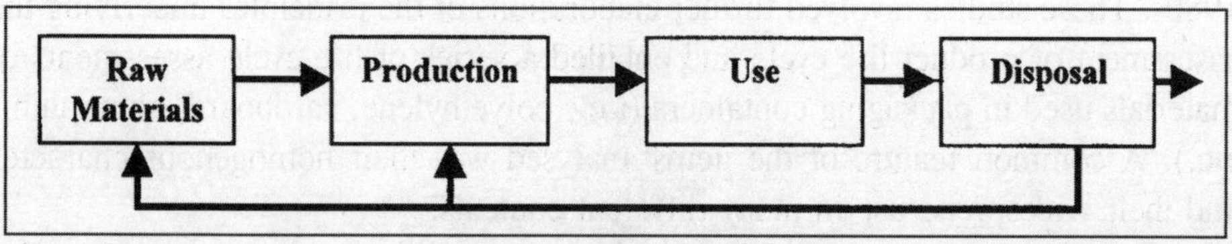

Figure 6.1 Product Life Cycle

The assessment process includes identifying and quantifying energy and materials used and wastes released to the environment, assessing their environmental impact and evaluating opportunities for improvement as illustrated in Figure 6.2.

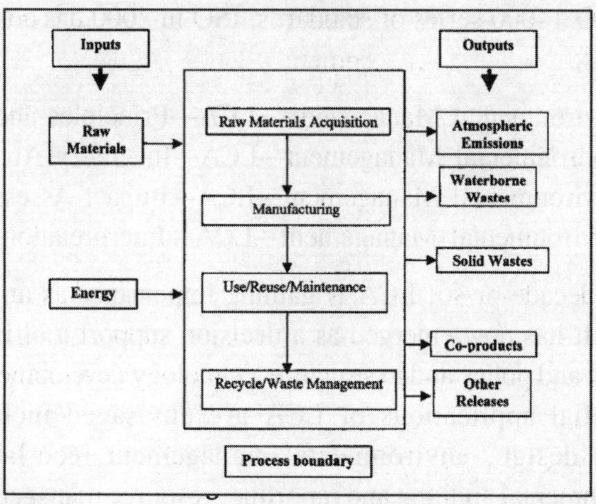

The unique feature of this type of assessment is its focus on the entire life cycle, rather than a single manufacturing step or environmental emission. The theory behind this approach is that operations occurring within a facility can also cause impacts outside the facility's gates that need to be considered when evaluating project alternatives.

The principles underlying an LCA were developed in the late 1960s for the evaluation of environmental impacts of products and were popularly known as resource and environmental profile analysis (REPA). Initially, it was used mainly on the consumption of energy and other resources. Knowledge of environmentally damaging releases and actions and the estimation of their effects, was too rudimentary at that time to allow a quantitative treatment of the environmental impacts of the product life cycle.

Assessments of product life cycle experienced a renaissance through studies of the environmental loadings and potential impacts of beverage containers (e.g., beer cans, milk containers) performed in various European countries in the early 1980s. These studies involved further elaborations of the principles underlying the assessment of product life cycle and entailed a series of life cycle assessments of materials used in packaging containers (i.e., polyethylene, cardboard, aluminium, etc.). A common feature of the items analysed was their homogenous character and their widespread use in many different contexts.

The late 1980s and early 1990s have seen international attempts to standardise the principles underlying life cycle assessments and to develop codes of good conduct in this field. The list of products that have been subjected to LCA has grown quickly and now includes more complex products such as paints, insulation materials, window frames, refrigerators, hotplates, television sets, etc., as well as the entire service systems or technologies such as electricity production.

As a part of ISO 14000 series of standards, ISO in 2000 has come out with the following standards:

- ISO 14040: Environmental Management—LCA—Principles and Framework.
- ISO 14041: Environmental Management—LCA—Inventory Analysis.
- ISO 14042: Environmental Management—LCA—Impact Assessment.
- ISO 14043: Environmental Management—LCA—Interpretation.

Since the last decade or so, LCA is gaining importance as an environmental management tool. It has now emerged as a decision support tool in such areas as business, regulation and policy and to structure technology development in a coherent way. Many potential applications of LCA are envisaged including product improvement and design, environmental management, eco-labelling, green accounting, environmental auditing and reporting, resource management, definition of best available technology (BAT), product policy, strategic industrial planning, strategic environmental policy development, etc. As a general concept, the life cycle approach aims to support the overall goal of sustainability.

LEARNING ACTIVITY 6.1

State the basic concept of a life cycle assessment.

Note:

a) Write your answer in the space given below.
b) Check your answer with the one given at the end of this book.

6.2 STAGES IN PRODUCT LCA

Figure 6.3 gives a diagrammatic representation of the stages in the life cycle of a product.

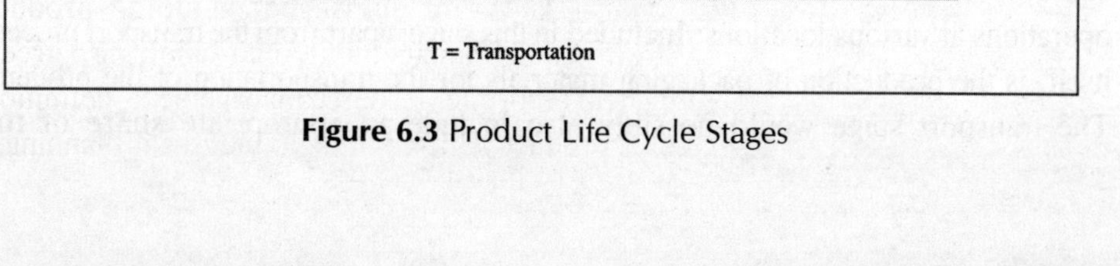

Figure 6.3 Product Life Cycle Stages

Note that Figure 6.3 complements Figures 6.1 and 6.2. As depicted in these figures, each stage of the life cycle receives materials and energy as inputs and produces:

- outputs of material or products to subsequent life cycle stages;
- emissions to the environment.

There are a number of issues associated with the life cycle stages, and we will touch upon some of these, next.

Extraction of Raw Materials

This stage in the life cycle includes the extraction of all materials involved in the entire life cycle of the product. Typical examples of activities included in this stage are forest logging and crop harvesting, fishing and mining of ores and minerals. The inventory for the extraction of raw materials should include raw materials for the production of the machinery (i.e., capital equipment) involved in manufacturing the product and other stages of the product life cycle. Raw materials used in the production of electricity and energy used in the different life stages of the product should also be considered. Collecting data for the raw materials extraction stage may prove to be a complex task. It may also lead into iterative processes such as assessing the inputs and outputs related to extraction of the raw materials that is used in the production of end products. Often, the most serious environmental problems of the product life cycle are associated with this first stage. It is a common error to leave out parts of the raw materials stage from the LCA. Essentially, the decision of what to include or exclude in the LCA should be based on a sensitivity analysis.

Manufacture of a Product

The manufacturing stage encompasses all the processes involved in the conversion of raw materials into the products considered in the LCA. Apart from the manufacturing processes at the plant where the product is made, this stage takes into account production of ancillary materials, chemicals and specific or general components at other plants, no matter where they are.

Transportation

As is indicated in Figure 6.3, transportation is really not a single life stage in itself. Rather, it is an integral part of all stages of the life cycle. Transportation could be characterised as conveyance of materials or energy between different operations at various locations. Included in this stage, apart from the transport process itself, is the production of packaging materials for the transportation of the product. The transport stage would possibly also include an appropriate share of the environmental loadings and consumptions associated with the construction and maintenance of the transport system, whether this is road, rail, water or air transportation.

Use of Product

The use-stage of the product occurs when it is put in service and operated over its useful life. This begins after the distribution of the product and ends when the product is used up or discarded to the waste management system. Included in the use-stage are releases and resource consumptions created by the use or maintenance of the product.

Waste Management

Wastes are generated in each phase of the life cycle, and they need to be properly managed to protect the environment. The management of wastes may involve alternative processes such as the following:

(i) **Reuse:** This means the use of the product or parts thereof in new units of the same product or in different products.
(ii) **Recycling:** This means the use of materials in the product for manufacture of the same or other products.
(iii) **Incineration:** This refers to the combustion of the product, generating heat that may be used for electricity production or heating.
(iv) **Composting:** This refers to the microbial degradation of biological materials yielding compost for improvement of agricultural soils.
(v) **Waste water treatment:** This refers to the organic matter degradation and nutrients removal from sewage water, creating sludge that is deposited on agricultural land.
(vi) **Land filling:** This means the deposition of the product in landfills.

Each form of waste treatment mentioned above may be considered as processing of waste associated with a certain consumption of resources. This results in various releases into the environment, and the possible generation of energy or materials that will be an input to the manufacturing process of this product or of other products.

Before you read any further, it is a good idea to look at the Course **Municipal Solid Waste Management** in which we discussed the different waste management processes in detail.

As implied, the LCA of products is an important environmental management tool. However, as with every tool, difficulties do arise with LCA too. In Section 6.3, we will discuss the problems and the questions that have to be taken into consideration, while analysing the life cycle of a product.

LEARNING ACTIVITY 6.2

Which of the materials described in section 6.2 (Product Life Cycle Stages) would you consider to be the least harmful to the external environment and the working environment?

Note:

a) Write your answer in the space given below.

b) Check your answer with the one given at the end of this book.

6.3 A CODE OF GOOD CONDUCT FOR LCA

As mentioned earlier, LCA emerged during the last decade as a tool to provide an objective assessment of the total environmental impact associated with a product through its entire life cycle. In several countries, LCA is considered the primary tool by which environmental impacts of products should be regulated by government authorities. However, problems do exist. Let us look at some of the major LCA limitations below, as revealed in various studies:

(i) **Data quality:** In a manufacturer-sponsored study to compare a product with its alternatives, the consultants performing the LCA were able to get a very detailed and current data from the manufacturer for the processes involved in the production of that product. However, they had to depend on secondary data from the literature or earlier studies with regard to the production processes concerning the alternative products. Obviously, comparative studies on the basis of the secondary data tend to lack credibility.

(ii) **Life cycle boundaries:** A Dutch study excluded the production of several raw materials, including crude oil, for the polycarbonate production. The German study did not include emissions from the energy production associated with the life cycle of the milk containers. Most studies did not consider the working environment.

(iii) **Country-specific technology types:** An example is the LCA conducted on the production of electricity used in the product life cycle. In a Swedish study, the electricity production was based on nuclear power and hydroelectricity, while a Swiss study, based on a US energy scenario since 1972, used coal as the energy source. This should explain the difference in the emissions found in the two studies. In addition, as the waste processing systems contribute to the nature and level of emissions, it is important to examine these systems as well. For example, if the waste is incinerated and the combustion energy is used for electricity production, the Dutch study may tend to favour carton containers over recyclable polycarbonate bottles.

(iv) **Evaluation stage priorities:** In the cases where alternatives do not have particular advantage over one another, the priorities in the evaluation stage become decisive for the outcome. A recommendation of the Danish study (referred to at (ii) above) was based on an evaluation stage choice between reduced consumption of raw materials and water, but on the releases of dichloromethane into the working environment (for polycarbonate bottles), and higher energy consumption and loading of bio-accumulating and possibly carcinogenic chlorinated compounds in the waste from production (for the milk cartons).

Considering the versatility and many diverse possible applications of the LCA tool, it may be difficult to obtain reproducible and consistent results through standardisation, without losing the necessary flexibility for adaptation to particular cases being studied. Given this scenario, it is important that we include the following factors in a code of good conduct for LCA.

(v) **Definition:** This involves finding answers to such questions as: Is the purpose of the study explicitly defined? Is it meant for internal company use or for public use? If the study is intended for public use, has it been peer reviewed? Is it clearly stated for whom the study is performed and by whom it is sponsored? Is the definition of the functional unit appropriate?, etc.

(vi) **Delimitation of system under study:** This involves finding answers to such questions as: Is there an explicit and clear delimitation of the system under study? Is the life cycle described in detail, stage by stage? Are the life cycle description and process tree plausible? Do they describe the real world system in a realistic way? Does the study include the extraction of raw materials? Does the study include the production of electricity? What production scenario is used? Is it appropriate? Does the study include the manufacture of real capital for all life cycle stages? If the study has precluded capital, is this omission substantiated? Do you find the omission reasonable? Is the disposal stage covered by the study?, etc.

(vii) **Inventory:** This involves finding answers to such questions as: Does the inventory cover all processes of the process tree? Is there a reference to the source of every piece of data in the inventory? Is the data quality appropriate, i.e., primary and recent data for all the important processes of the life cycle? Do the data describe a relevant technological level of the processes? Is the use of data of lower quality or omission of processes from the inventory based on sensitivity analyses? Is the quality of the data used for the compared alternatives similar? Are data on impacts to the working environment present in the inventory? Is it acceptable that they be left out in the considered study?, etc.

(viii) **Impact assessment:** This involves finding answers to such questions as: Has any impact assessment been performed? Does it consider all the important environmental effect types? Does it consider resource and working environment issues?, etc.

(ix) **Evaluation:** This involves finding answers to such questions as: Are the subjective steps of the evaluation separated from the objective ones? If not, is the evaluation transparent? Are the priorities clear? Are the conclusions of the assessment clear? Are the assumptions underlying the weighting explained, and do you agree with them?, etc.

6.4 PROCEDURE FOR LCA

Four steps are involved in carrying out an LCA, and these are:

(i) Definition of scope, goals, and delimitation of the life cycle.
(ii) Preparation of an inventory.
(iii) Assessment of impact of environmental loadings in terms of environmental profiles.
(iv) Evaluation of environmental profiles according to the defined goals.

In Subsections 6.4.1 to 6.4.4 we will discuss these steps in detail.

6.4.1 Defining the Goal and Scope

The goal of an LCA study must unambiguously state the intended application, including the reasons for carrying out the study and the intended audience. The statement of goal must also indicate the intended use of the results and users of the results. The practitioner, who has to reach the goal, needs to understand the detailed purpose of the study in order to make proper decisions throughout the study. Examples of goals of life cycle assessments are:

- To compare two or more different products fulfilling the same function with the purpose of using the information in marketing of the products or regulating the use of the products.
- To identify improvement possibilities in further development of existing products or in innovation and design of new products.
- To identify areas, steps, etc., in the life cycle of a product to meet the eco-labelling criteria.

Transparency is essential for all kinds of LCA studies. The target group of the LCA study is also important to have in mind the choice of reporting method. The goal can be redefined as a result of the findings throughout the study.

While the goal definition determines the level of sophistication of the study and the requirements to reporting, the definition of the scope of an LCA sets the borders of the assessment, i.e., what may be included in the system and what detailed assessment methods are to be used.

In defining the scope of an LCA study, the following elements must be considered and clearly described:

- The functions of the system, or in the case of comparative studies, systems.
- The functional unit.
- The system to be studied.
- The system boundaries.
- The allocation procedures.
- The types of impact and the methodology of impact assessment and subsequent interpretation to be used.
- Data requirements.
- Assumptions.
- Limitations.
- The initial data quality requirements.
- The type of critical review, if any.
- The type and format of the report required for the study.

The scope should be sufficiently well defined to ensure that the breadth, the depth and the detail of the study are compatible and sufficient to address the stated goal. LCA is an iterative technique. Therefore, the scope of the study may need to be modified while the study is being conducted as additional information is collected.

Let us now describe some of the elements mentioned above to define the scope of an LCA.

Functional Unit

Definition of the functional unit or performance characteristics is the foundation of an LCA because the functional unit sets the scale for comparison of two or more products including improvement to one product (system). All data collected in the inventory phase will be related to the functional unit. When comparing different products fulfilling the same function, definition of the functional unit is of particular importance. The three aspects that have to be taken into account when defining a functional unit are the efficiency of the product, the durability of the product and the performance quality standard (Lindfors et al., 1995).

When performing an assessment of more complicated systems, e.g., multi-functional systems, special attention has to be paid to by-products. Waste treatment systems are an example of processes with different outputs (e.g., energy, a fertiliser). When comparing different systems, inclusion of the produced amount of energy and fertiliser is an example of handling of different by-products in the definition of the functional unit.

System Boundaries

The system boundaries define the processes/operation (e.g., manufacturing, transport and waste management processes) and the inputs and outputs to be taken into account in the LCA. The input can be the overall input to a production as well as input to a single process, and the same is true for the output. The definition of system boundaries is a quite subjective operation and includes geographical boundaries, life cycle boundaries (i.e., limitations in the life cycle) and boundaries between the technosphere and biosphere. Due to the subjectivity of definition of system boundaries, transparency of the defining process and the assumptions are extremely important. Note that wastewater treatment is an example of a process that is often omitted when defining the system boundaries.

Data Quality

The quality of the data used in the life cycle inventory is naturally reflected in the quality of the final LCA. The data quality can be described and assessed in different ways. It is important that the data quality is described and assessed in systematic ways that allows others to understand and control the actual data quality. Further descriptions, which define the nature of the data collected from specific sites and that from published sources and indicate whether the data should be measured, calculated or estimated, should also be considered for maintaining data quality.

In all studies, however, the following additional data quality indicators are to be taken into consideration at a certain level of detail, depending on goal and scope definition:

- **Precision:** That is, the measure of variability of the data values for each category expressed (e.g., variance).
- **Completeness:** That is, the percentage of locations reporting primary data from the potential number in existence for each data category in a unit process.
- **Representativeness:** This means the qualitative assessment of the degree to which the data set reflects the true population of interest (i.e., geographic, time period and technology coverage).
- **Reproducibility:** This means the qualitative assessment of the extent to which formation about the methodology and data values allows an independent practitioner to reproduce the results reported in the study.
- **Consistency:** This means the qualitative assessment of how uniformly the study methodology is applied to the various components of the analysis.

Critical Review Process

The purpose of the critical review process is to ensure the quality of the life cycle assessment. The review can be either internal or external. This may also involve the interested parties as defined within the goal and scoping definition. The critical review process ensures that:

- the methods used to carry out the LCA are consistent with the international standard and are scientifically and technically valid;
- the data used are appropriate and reasonable in relation to the goal of the study;
- the interpretations reflect the limitations identified and the goal of the study;
- the study report is transparent and consistent.

6.4.2 Analysing the Inventory

Inventory analysis is the second phase in an LCA, consisting of issues such as data collection, refining system boundaries, calculation, verification of data, relating data to the specific system and allocation.

We will describe each of these issues, next.

Data Collection

The inventory analysis includes collection and treatment of data to be used in preparation of a material consumption, waste and emission profile for the phases in each life cycle. The data can be both site-specific (e.g., from specific companies, specific areas and specific countries) and general sources (e.g., trade organisations, public surveys, etc.). The data have to be collected from all single processes in the life cycle. The quantitative data are important in comparisons of processes or materials, but often the quantitative data are missing or the quality is poor (too old or not technically representative, etc.). However, a more descriptive qualitative data can be used for environmental aspects or single steps in the life cycle that cannot be quantified. This can be used when the goal and scope definitions allow a non-quantitative description of the conditions.

Data collection is often the most work intensive part of a life cycle assessment, especially if site-specific data are required for all the single processes in the life cycle. In many cases, average data from the literature (often previous investigations of the same or similar products or materials) or data from trade organisations are used. The average data can be used in the conception or simplified LCA to get a first impression of the potential inputs and outputs from producing specific materials. When doing a detailed LCA, site-specific data is preferred. Note that since average data are often some years old, they may not represent the latest in technological development.

Refining System Boundaries

The system boundaries are defined as part of the scope definition procedure. After the initial data collection, the system boundaries can be refined, i.e., as a result of decisions of exclusion of life stages or sub-systems, exclusion of material flows or inclusion of new unit processes shown to be significant, according to the sensitivity analysis. The results of this refining process and the sensitivity analysis are to be documented. This analysis serves to limit the subsequent data handling to those inputs and output data, which are determined to be significant to the goal of the LCA study.

Calculation Procedures

No formal demand exists for calculation in life cycle assessment except the described demands for allocation procedures. Due to the amount of data, it is recommended, as a minimum, to develop a spreadsheet for the specific purpose. The appropriate programme can be chosen, depending on the kind and amount of data to be handled.

Validation of Data

The validation of data has to be conducted during the data collection process in order to improve the overall data quality. Systematic data validation may point out the areas where data quality must be improved or data must be found in the similar processes. During the process of data collection, a permanent and iterative check on data validity should be conducted. Validation may involve establishing mass balances, energy balances and/or comparative analyses of emission factors. Obvious anomalies in the data appearing from such validations will result in alternative data values complying with the data quality requirements.

For each data category and for each reporting location where missing data are identified, the treatment of the missing data should result in an acceptable reported data value, a *zero* data value if justified and a calculated value, based on the reported value from unit processes employing similar technology.

Relating Data

The fundamental input and output data are often delivered from industry in arbitrary units, e.g., energy consumption as MJ/machine/week emissions to the sewage system as mg metals/litre wastewater. The specific machine or wastewater stream is rarely connected to the production of the considered product alone but often to a number of similar products or perhaps to the whole production activity.

For each unit process, an appropriate reference flow has to be determined (e.g., one kilogram of material, one mega joule for energy, etc.). The quantitative input and output data of the unit process shall be calculated in relation to this reference flow. Based on the refined flow chart and systems boundary, unit processes are interconnected to allow calculations of the complete system. The calculation should result in all system input and output data being referenced to the functional unit. Care should be taken when aggregating the inputs and the outputs in the product system. The level of aggregation should be sufficient to satisfy the goal of the study.

Allocation and Recycling

When performing a life cycle assessment of a complex system, it may not be possible to handle all the impacts and outputs inside the system boundaries. This problem can be solved through either of the following ways:

- expanding the system boundaries to include all the inputs and outputs;
- allocating the relevant environmental impacts to the studied system.

When avoiding allocation by expanding the system boundaries, there is a risk of making the system too complex. The data collection, impact assessment and interpretation can then become too expensive and unrealistic in terms of time and money. Allocation may be a better alternative, if an appropriate method can be found for solving the actual problem. Allocation can be necessary when dealing with:

- *multi-output black box* processes, i.e., when more than one product is produced and some of those product flows are crossing the system boundaries;
- multi-input processes, such as waste treatment, where a strict quantitative causality between inputs and emissions seldom exists;

- open-loop recycling, where a waste material leaving the system boundaries is used as a raw material by another system, outside the boundaries of the studied system.

On the basis of the principles presented above, the following descending order of allocation procedures is generally recommended:

- Wherever possible, allocation should be avoided or minimised. This may be achieved by subdividing the unit process into two or more sub-processes, some of which can be excluded from the system under study. Transport and materials handling are examples of processes, which can sometimes be partitioned in this way.
- Where allocation cannot be avoided, the system inputs and outputs should be partitioned among their different products or functions in such a way as to reflect the underlying physical relationship among them. That is to say, they must reflect the way in which the inputs and outputs are changed by quantitative changes in the products or functions delivered by the system.
- Where physical relationship cannot be established or used as the basis for allocation, the inputs should be allocated between the products and functions in such a way as to reflect economic relationships between them. For example, burdens might be allocated between co-products in proportion to the economic value of the products.

6.4.3 Assessing Environmental Impact

Impact assessment involves category definition, classification, characterisation and valuation/weighting. Let us now discuss each of these elements.

Category Definition

The life cycle assessment involves, as a first element, the definition of the impact categories to be considered (ISO, 1997). This is a follow-up of the decisions made in the goal and scoping phase. Based on the type of information collected in the inventory phase, however, the boundaries defined in the goal and scoping may be redefined.

The impact categories are selected in order to describe the impacts caused by the products or product systems considered. The issues that need to be considered when selecting impact categories include the following:

- **Completeness**: This means that all environmental problems of relevance should be covered by the list.
- **Practicality**: This means that the list should not contain too many categories.

- **Independence**: This means that double counting should be avoided by choosing mutually independent impact categories.
- **Relation to the characterisation step**: This means that the chosen impact categories should be related to available characterisation methods (Lindfors et al., 1995).

The impact categories considered are abiotic resources, biotic resources, land use, global warming, stratospheric ozone depletion, ecotoxicological impacts, human toxicological impacts, photochemical oxidant formation, acidification, eutrophication and work environment.

Classification

The life cycle impact assessment includes, as a second element, classification of the inventory input and output data (ISO, 1997).

Classification is a qualitative step based on a scientific analysis of the relevant environmental processes. The classification has to assign the inventory input and output data to potential environmental impacts, i.e., impact categories. Some outputs contribute to more than one impact category, and therefore, they have to be mentioned twice. The resulting double counting is acceptable if the effects are independent of each other. However, double counting of different effects in the same effect chain (e.g., stratospheric ozone depletion and human toxicological effects, such as skin cancer) is not allowed.

The impact categories can be placed on a scale dividing the categories into four different space groups: global impacts, continental impacts, regional impacts and local impacts. The grouping is not mutually exclusive (e.g., environmental toxicity can be global, continental, regional as well as local). The impact categories are often related directly to *exposure* (e.g., global exposure leads to continental impacts). Some of the impact categories are strongly correlated with continental, regional or local conditions. Certain lakes in Scandinavia can be mentioned as examples of localities that are more predisposed to acidification than lakes in other parts of Europe. The *time* aspect is also important when considering certain impact categories (e.g., global warming and stratospheric ozone depletion with time horizons of 20 to 500 years).

Now, let us consider an example of classification. In the manufacture of refrigerators, many of the activities relating to the production process involve the combustion of fossil fuels (e.g., coal, oil and natural gas) for electricity or heat generation, and combustion causes various emissions that may be classified as given in Fig. 6.4.

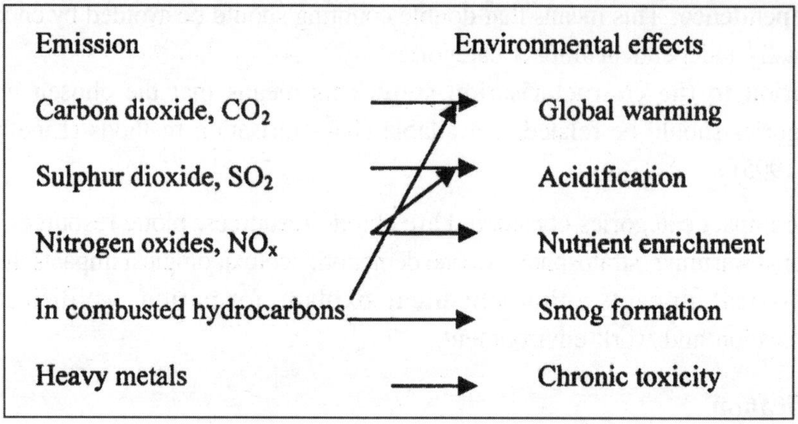

Figure 6.4 Classification of Emissions

Characterisation

The life cycle impact assessment includes, as a third element, the characterisation of the inventory data (ISO, 1997).

Characterisation is mainly a quantitative step, based on scientific analyses of the relevant environmental processes. It has to assign the relative contribution of each input and output to the selected impact categories. The potential contribution of each input and output to the environmental impacts has to be estimated. For some of the environmental impact categories, there is consensus about equivalency factors to be used in the estimation of the total impact (e.g., global warming potentials, ozone depletion potentials, etc.) while for some others there may not be any consensus (e.g., biotic resources, land uses, etc.).

Valuation/Weighting

Characterisation results in a quantitative statement on different impact categories (e.g., global warming, stratospheric ozone depletion and ecotoxicological effects). Comparison of these categories is not immediately possible. Therefore, the life cycle assessment includes, as a fourth element, a valuation/weighting of the impact categories against each other.

Weighting aims to rank, weight, or if possible, aggregate the results of different life cycle impact assessment categories in order to arrive at the relative importance of these different results. The weighting process is not technical, scientific or objective, as these various life cycle impact assessment results are not directly comparable (e.g., indicators for greenhouse gases or resource depletion). However, applying scientifically based analytical techniques may assist in weighting. The purpose of weighting is the following aspects:

- To express the relative preference of an organisation or group of stakeholders based on policies, goals or aims and personal or group opinions or beliefs common to the group.
- To ensure that process is visible, documentable and reportable.
- To establish the relative importance of the results is based on the state of knowledge about these issues.

Different institutions based on different principles have developed weighting approaches including the following:

- **Proxy approach:** In this approach, one or a combination of several quantitative measures is stated to be indicative of the total environmental impact. Energy consumption, material displacement and space consumption are examples of using this approach.
- **Technology abatement approach:** The possibility of reducing environmental burdens by using different technological abatement methods can be used to set a value on the specific environmental burden. This approach can be applied to inventory data as well as impact scores.
- **Monetarisation:** This approach is based on the premise that utilitarianism (values are measured by the aggregation of human preferences), willingness to pay/ accept is an adequate measure of preferences and the values of environmental quality can be substituted by other commodities. This approach can be applied to inventory data as well as impact scores.
- **Authorised goals or standards:** Environmental standards and quality targets as well as political reduction targets can be used to calculate critical volumes for emissions to air, water, soil or work environment. National or local authorities within a company can formulate the targets or standards.
- **Authoritative panels:** An authoritative panel can be made up of lay people, experts, representatives of governments, international bodies, etc. The credibility of a panel can be improved by using LCA experts from different societal groups as panellists, peer review sets of valuation criteria and the rules for their application, and a transparent ranking technique, and the documentation of the arguments leading to the final valuation (Volkwein et al., 1996).

In the impact assessment step of an LCA, the inventory of chemical and physical releases is translated into an environmental profile and the resources consumed are divided into renewable and non-renewable. Table 6.1 illustrates environmental releases and resource consumption.

Table 6.1 Quantitative Inventories of Environmental Releases and Resources Consumptions for the Production of 1 MJ Heat from Various Fossil Fuels

Emission type	*Unit*	*Coal*	*Heavy fuel*	*Natural gas*	*Wood*	*Straw*
Carbon dioxide, CO_2	g/MJ	96	80	56	98a	99a
Carbon monoxide, CO	mg/MJ	50	10	150	6000	4500
Sulphur dioxide, SO_2	mg/MJ	650	500	1	33	110
Nitrogen oxides, NO_x	mg/MJ	500	240	350	50	130
Volatile organic compounds, VOC	mg/MJ	15	7	15	1200	-
Polycyclic aromatic hydrocarbons, PAH	mg/MJ	2.0	0.25	0.13	5000	25
Dioxins, PCDD, PCDF	mg/MJ	0.04	-	0	-	0.3
Arsenic, As	mg/MJ	8	2	0.0	5	-
Cadmium, Cd	mg/MJ	0.9	0.7	0.0	10	-
Chromium, Cr	mg/MJ	4	1.2	0.0	50	-
Mercury, Hg	mg/MJ	1.7	0.1	0.0	1	0.0
Nickel, Ni	mg/MJ	6	400	0.0	-	-
Lead, Pb	mg/MJ	7.6	25	0.0	200	1.0
Particles	mg/MJ	25	1.3	0	45	20
Resource consumption	g/MJ	39	25	23	100	67

Note that 'a' in the table above indicates that the fossil fuel does not contribute to the increase in the CO_2 level of the atmosphere.

The environmental profile shows the potential contribution to all types of environmental effects that are considered important and undesirable. For each individual release listed in the inventory, its potential contributions are calculated and added to the overall contribution to the relevant effect type. Releases that do not contribute to any known effect type will not influence the environmental profile. The list is obviously subject to change due to developments in scientific knowledge, and, to a certain extent, political initiatives. The list should, therefore, be continually adjusted to accommodate such changes. Table 6.2 gives a list of the environmental effects that ought to be considered in an assessment.

Let us now work out Learning Activity 6.3, before we proceed any further.

Table 6.2 LCA: Environmental Effect Types*

Global	Global warming Stratospheric ozone depletion
Regional	Acidification Nutrient enrichment Photochemical ozone formation (smog) Chronic toxicity
Local	Acute toxicity Area degradation, including: • risks • noise • smell Physical disturbances, including: • soil erosion • deterioration in landscape quality • disturbance or destruction of habitats, and hence, ecosystems.

* *See also Case 6.2 (i.e. Section 6.2)*

LEARNING ACTIVITY 6.3

Table 6.1 gives a quantitative inventory of the most important environmental releases and resource consumptions for the production of 1 MJ heat, using the following fossil fuels: coal (coal powder, 1% sulphur content), heavy fuel (1% sulphur content), natural gas, or the biomass of wood or straw. For all fuel types except wood, the burners are equipped with filters for removing particles (equipping wood burners with particle filters would lower emissions considerably). No further treatment of flue gas is performed.

(i) Classify all the environmental releases into the different types of environmental effects that they will potentially contribute and compute the contribution. Classify the resources as renewable or non-renewable.

(ii) Which of the fuels do you find to be of the greatest environmental concern?

Note:

a) Write your answer in the space given below.

b) Check your answer with the one given at the end of this book.

Contribution to	Unit	Coal	Heavy Fuel	Natural Gas	Wood	Straw

6.4.4 Evaluating Environmental Profiles

Evaluation involves a comparison of the alternatives studied (e.g., products, materials and projects), represented by the environmental, resource and the working environment profiles. The objective of the evaluation is to decide which alternative causes the least environmental damage from a life cycle point of view.

If one alternative is better in all the aspects considered, the evaluation is straightforward. In fact, this is often the case when a product is reviewed first with the purpose of reducing its environmental impacts. However, the evaluation situation gets complex when one alternative is better than the rest in some aspects, while another alternative is better in other aspects. This creates the need to give priorities to the different aspects of the loading profiles. For example, we may ask the questions:

- Which is more important—improving performance in the external environment or in the working environment?
- Is reducing the life cycle contribution to ozone depletion more important than diminishing the release of chemicals causing chronic toxicity?

Clearly, no objective answer can be given to questions like these that involve judgement of fundamentally incomparable items. The setting of the priorities involved in the evaluation step inevitably introduces subjective decisions into the LCA. It should be kept in mind that all the work conducted in LCA—steps leading up to the evaluation must be based on objective and scientific principles. The subjective decisions are hence confined to this last step.

When the inventory is complete and all the emissions listed are classified into environmental impact classes, and converted into common effect units, where possible, the result is an environmental profile for each of the alternatives examined, as shown in Fig. 6.5.

Comparing the profiles, you may find that Alternative 1 contributes less than Alternative 2 to global warming, acidification and nutrient enrichment, while Alternative 2 contributes less to ozone depletion and acute toxicity.

It may be very difficult to weigh between different classes of environmental impacts. But, it will help, if you know which of the differences in contribution to the environmental effects are important and which are not. An example of the kind of question you may have to answer is: How does 15 kg of CO_2 equivalents compare to 3 kg of CFC-11 equivalents, or to 5 kg of SO_2 equivalents?

To help in the comparisons of this nature, each of the contribution in the environmental profile may be divided by the emission from a suitable common emission scenario. This scenario may, for instance, be the environmental profile of the total emissions of your state or country. In this case, the resulting normalised environmental profile shown in Fig. 6.6 indicates the size of the fraction of national emissions due to each of the environmental effect contributions.

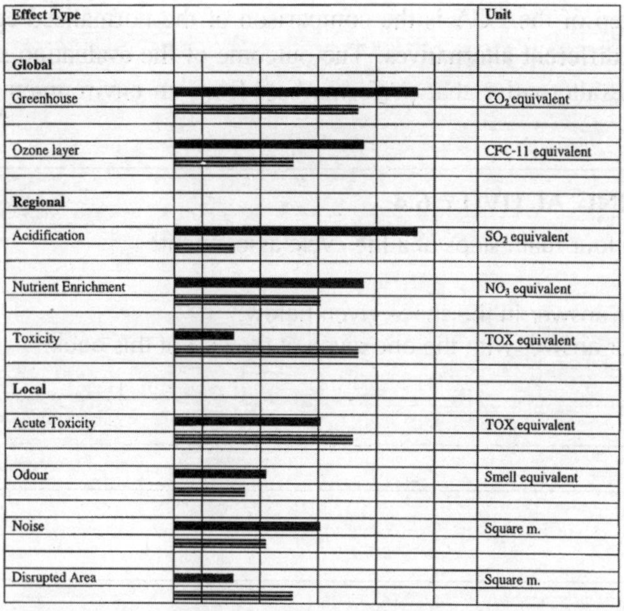

Figure 6.5 Environmental Profile of Two Product Alternatives

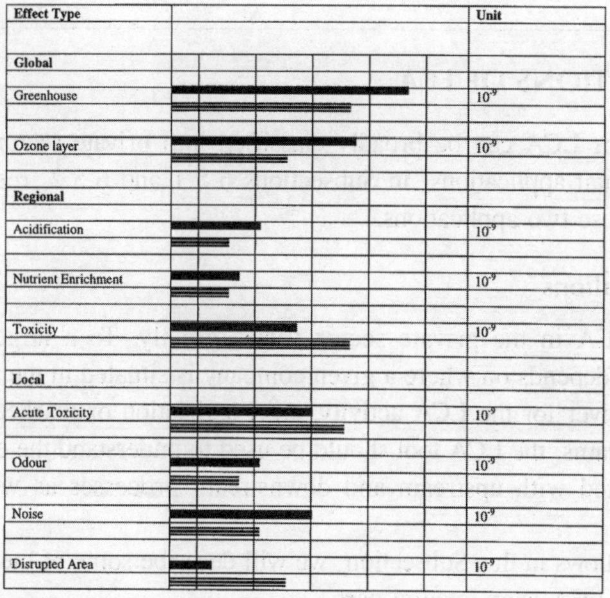

Figure 6.6 Normalised Environmental Profiles of Fig. 6.5

The normalisation step is of particular relevance for comparing alternatives with complicated environmental profiles, i.e., the many different emissions that contribute to different impacts on the environment.

The last step of the LCA is the comparison of the normalised environmental profiles of the different alternatives. The outcome of the evaluation should be the selection of the alternative that performs best from an environmental protection viewpoint.

LEARNING ACTIVITY 6.4

What are the four main steps of a life cycle assessment?

Note:

a) Write your answer in the space given below.

b) Check your answer with the one given at the end of this book.

6.5 DIFFERENT APPLICATIONS OF LCA

Applications for LCA can be broadly classified into private sector applications and governmental applications. In Subsections 6.5.1 and 6.5.2, respectively, we will discuss these two applications.

6.5.1 Private Sector Applications

The use of LCA in the private sector varies greatly. To a large extent, this differentiation depends on where a given company is situated in the product chain and the key driver for the LCA activity, e.g., legislation or market competition. For business teams, the LCA tool should be used to understand the environmental issues associated with upstream and downstream processes as well as on-site processes.

In what follows in this Subsection, we will describe some of the concepts and tools in which LCA is an integral part.

Product Development

Using LCA in product development is an obvious choice, as a large part of the future environmental impacts of a product (system) is determined by the design and construction phase. By incorporating LCA in the design phase, companies have the possibility of avoiding or minimising foreseeable impacts compromising the overall quality of the product.

Product development may follow different concepts and routes. Figure 6.7 shows some of the common phases of most development methodologies.

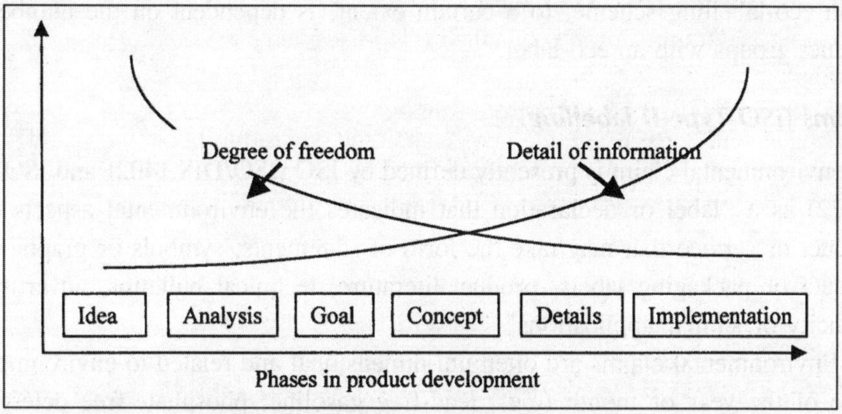

Figure 6.7 Relationships Between the Designer's Degree of Freedom and the Level of Information
Adapted from Hanssen, 1995.

In the idea-phase, there is almost an unlimited number of possibilities with respect to design, choice of materials, function, etc. The number of options decreases with the development process and changes to the final product and the necessary productive tools often require a whole new development process. It is, therefore, necessary that relevant environmental tools are available and used as early as possible in the development process. For simple products, e.g., packaging, it is possible to apply a detailed and quantitative LCA, since the information on most of the commonly used materials is now available.

Marketing

Marketing is the traditional way of communicating product properties and capabilities, which are consistent with the consumer's expectations and demands. As the level of environmental consciousness is increasing, the consumer pays more attention to the environmental properties of goods and services. There are four different kinds of environmental marketing. We will describe these, next.

Environmental Labelling (ISO Type-I labelling)

An environmental label (eco-label) can be seen as a seal of approval for environmentally benign products and can, therefore, be attractive for marketing purposes. Eco-label, at the same time, conveys information to the consumer in a simple but not in an objective way, enabling individuals to include environmental concerns in their own decisions along with considerations for economy and quality.

The general objective of national and supranational eco-labelling scheme is to make products with less environmental impacts visible to the consumer. The success of an eco-labelling scheme, to a certain extent, is dependent on the number of product groups with an eco-label.

Environmental Claims (ISO Type-II labelling)

An environmental claim is presently defined by ISO (ISO/DIS 14021 and ISO/CD 14022) as a "label or declaration that indicates the environmental aspects of a product or service that may take the form of statements, symbols or graphics on product or packaging labels, product literature, technical bulletins, advertising, publicity or similar applications".

Environmental claims are often uni-dimensional and related to environmental issue of the year or month (e.g., lead-free gasoline, phosphate-free detergent, CFC-free hair spray, etc.). Very few of such environmental claims are based on an LCA and, in many cases, focus on irrelevant issues, while failing to address those issues, which are important in a life cycle perspective.

Environmental Declarations (ISO Type-III labelling)

Environmental declarations may be a tool in eco-marketing to transfer the results from a life cycle investigation of a product (either as a life cycle inventory or a life cycle assessment) to the individual decision-making process of a consumer. The general idea is to give a graphic presentation of a pre-set number of environmental impacts, e.g., by using a bar diagram. Two or more environmental declarations can, on the other hand, illustrate the actual difference between the products and, for environmentally conscious consumers, this can be a valuable supplement to choose between labelled products.

The use of LCA is, thus, a prerequisite for environmental declarations. Standardisation efforts have been initiated by ISO and includes requirements on methodology, transparency, external review, comparative assertions, labelling components, administrative guidelines and procedures governing the accreditation and conduct of type III labelling practitioners.

Organisational Marketing

The classical marketing of environmental performance has mainly been oriented towards products. However, with the increasing number of companies being certified according to ISO 14001, EMAS or BS 7750, some marketing initiatives are being directed towards the environmental capabilities of the companies per se.

As organisations implement the necessary policies for certification, they also encourage formalising the implementation of LCA procedures and life cycle.

Strategic Business Planning

Integration of environmental aspects in strategic business planning is becoming a common feature in many companies like the handling of environmental management schemes such as EMAS (Environmental Management and Auditing Scheme) and ISO 14001 standards. But, many companies still handle the issues on a case-to-case basis.

There are several motivating factors behind the decision to integrate environmental issues, many of which are interrelated. These include, for example, consumer demands, compliance with legislation, community needs for environmental improvement, security of supply and product and market opportunities.

The environmental performance is, thus, changing from being a mandatory property of many products (i.e., all regulatory requirements shall be fulfilled) to being a strong positioning property on the market. An LCA can be used both in relation to existing products and to identify market segments to be opened for environmentally benign products. The LCA information can provide decision-makers with an understanding of the environmental pros and cons of their products and services.

6.5.2 Governmental Applications

Sustainable development has been included as a major item on most of the governmental agendas since the 1992 Rio summit. Although a precise definition of sustainable development has not been given, it is obvious that LCA must be used to ensure that actions towards a more sustainable future will have the desired effect. An LCA, as a specific tool, can ensure this in some cases, while LCA, as an approach or as a strategic tool, can give directions but not the whole answer and must, therefore, be applied along with other tools such as risk assessment, environmental impact assessment, cost benefit analysis and others. The main governmental applications include product oriented policy, deposit-refund schemes, including waste management policies, subsidies and taxation, and general (process-oriented) policies.

SUMMARY

In this Unit, we discussed life cycle assessment (LCA) as an environmental management tool. After tracing the evolution of LCA and describing the basic stages of a product life cycle, we identified the elements that contribute to the making of the code of good conduct for an LCA study. We then explained the steps such as goal setting, inventory analysis, impact assessment and profile evaluation involved in the LCA process. We close the Unit by discussing LCA applications both in private and public sectors.

REFERENCES

ISO,1997. Environmental management - Life cycle assessment - Life cycle impact assessment. ISO/CD 14 042.1, 1997.01.15.

Lindfors L-G, Christiansen K, Hoffman L, Virtanen Y, Juntilla V, Hanssen O-J, Rönning A, Ekvall T, Finnveden G., 1995. Nordic Guidelines on Life-Cycle Assessment. Copenhagen: Nordic Council of Ministers

Volkwein St., Klöpffer W., 1996. The Valuation Step in LCA: Part I: General Principles, *Int J LCA* 1(1): 36-9 (1996)

Hanssen O.J., 1995. Sustainable management of product systems - an integrated method for life cycle assessment and management of product systems. Oestfold Research Foundation,.

SUGGESTED READINGS

Nierynck, E. 1993. Life Cycle Assessment. *In:* Environmental Management, Volume 3. Instruments form Implementation, B. Nath, L. Hens and D. Devuys (eds). VUB University Press, Brussels.

OECD. 1994. *Life-Cycle Management and Trade*. OECD, Paris.

SETAC. 1993. Guidelines for Life-Cycle Assessment: A 'Code of Practice'. From the SETAC Workshop held at Sesimbra, Portugal, 31 March to 3 April, 1993. Society of Environmental Toxicology and Chemistry.

Weidema, B.P. (ed). 1993. Environmental Assessment of Products: A Texbook on Life Cycle Assessment. UETP-EEE The Finnish Association of Graduate Engineers, Helsinki, Finland.

CASE 6.1: SOFTWARES FOR PERFORMANCE OF LIFE CYCLE ASSESSMENT

Name	ECOINVENT (Environmental Life Cycle Inventories of Energy Systems)
Contact Person	Mr. Rolf Frischknecht Tel: 41-1-632 5795 Fax: 41-1-632 1283
Organisation	Energy-Materials-Environment Group Institute for Energy Technology ETH Zentrum-UNL CH-8092 Zurich Switzerland
Homepage	*http://www.esu.les.iet.mavt.ethz.ch/esu/Ecoinvent.html* *http://jesus.ethz.ch/ecoinven/ifc_dt/get/eco_get.html* *http://www.worldbank.org/html/fpd/em/eminfo/EA/methods/cstecoit.htm*
Name	BUWAL 250
Contact Person	
Organisation	ETH Swiss Federal Institute of Technology Zurich Institute of Process Engineering and Cryogenics ETH Zentrum CH-8092 Zurich fax: +41-1-632 32 79 tel: +41-1-632 32 79 SVI-Swiss Packaging Institute Bruckfeldstrasse 18 Postfach 3000 Bern 26 tel: +41-31-3023003 fax: +41-31-3023047
Homepage	*http://www.ethz.ch*
Name	Eco-profiles of the European plastics industry
Contact Person	
Organisation	APME-Association of Plastics Manufacturers in Europe
Homepage	*http://www.apme.org/index.htm*
Name	SimaPro
Contact Person	
Organisation	PRé Consultants BV Plotterweg 12 3821 BB Amersfoort The Netherlands Telephone: +31 33 4555022 Fax: +31 33 4555024 E-mail: info@pre.nl
Homepage	*http://www.pre.nl/index.html*

Name	Boustead model
Contact Person	Dr. Ian Boustead phone: +44-403-864 561 fax: +44-403-865 284
Organisation	Dr. Boustead Boustead Consulting 2 Black Cottages, Worthing Road, West Grinstead, Horsham, West Sussex, Great Britain RH13 7BD
Name	TEAM/DEAM
Contact Person	
Organisation	Ecobilan Ecobalance 15204 OMEGA DRIVE, SUITE 220 ROCKVILLE, MARYLAND 20850
Homepage	Ecobalance: *http://ourworld.compuserve.com/homepages/Ecobalance/BR0.htm*
Name	LCAD
Contact Person	Kenneth K. Humphreys Pacific Northwest National Laboratory Tel: 509-372-4279 Fax: 509-372-4370 e-mail: kk_humphreys@pnl.gov Bruce W. Vigon Battelle Columbus Operations Tel: 614-424-4463 Fax: 614-424-304 e-mail: vigonb@battelle.org
Organisation	Battelle Memorial Institute 505 King Avenue Columbus, Ohio 43201 Phone (614) 424-6424 Fax (614) 424-3404
Homepage	*http://www.estd.battelle.org/sehsm/lca/software.html*
Name	IDEMAT
Contact Person	J.A.M. Remmerswaal Phone: 31-15-278-5041 Fax: 31-15-278-2956 E-mail: j.a.m.remmerswaal@io.tudelft.nl
Organisation	Delft University of Technology Faculty of Industrial Design Engineering Department of New Product Development Section for Environmental Product Development Jaffalaan 9 2628 BX Delft The Netherlands

Homepage	*http://www.io.tudelft.nl/research/mpo/* *http://www.io.tudelft.nl/research/mpo/idemat/idemat.htm*
Data	Database for designers. Data for a number of different materials, processes and components. Data given for different environmental themes, and computed Eco-indicator and EPS-indicator.
Name	R & M Steel Company
Websites	http://www.rmsteel.com http://www.aviationbuildingsystem.com
CONTACT	P.O. Box 580 , Caldwell, Idaho 83606

CASE 6.2: ENVIRONMENTAL EFFECTS TO BE CONSIDERED IN LCA

Global Warming

This is caused by gases that absorb infra-red radiation (heat) irradiated from the earth. Increasing their concentrations in the atmosphere may cause the global atmospheric temperature to increase, and this may, through increased melting of polar ice caps and heat-expansion of water in the oceans, lead to rising sea levels and serious climatic changes. The range and the ecological and socio-economic consequences of these effects are impossible to predict at present.

Carbon dioxide, CO_2, is the most prominent of the global warming gases, contributing about 60% of the total anthropogenic (human-made) atmospheric content of global warming gases. Other important gases are methane, CH_4, contributing 15% (the main anthropogenic sources are stocks of cattle and cultivation of rice), dinitrogenoxide (or nitrous oxide), N_2O, with 4% (combustion processes, fertilised agricultural soils) and the halocarbons group, with around 10% (CFCs, HCFCs and halons, all anthropogenic in origin and manufactured for use as solvents, aerosol propellants, heat transmitters, insulators and fire extinguishers).

For carbon dioxide, it is important in the inventory to distinguish between activities that cause a net increase in the atmospheric content (such as combustion of the fossil fuels, oil, coal and gas, or their derivatives such as plastics) and other activities where the CO, CO_2 contribution would reach the atmosphere anyway, through the biogeochemical carbon cycle. The latter includes the burning of biological materials that would otherwise have decayed naturally and released their carbon content as CO_2 to the atmosphere. However, the clearance and burning of forests to create agricultural land is an exception from this rule, since the activity causes a net reduction in the global biomass and thus a net increase in the atmospheric content of CO_2.

Stratospheric Ozone Depletion

The stratosphere is one of the higher strata of the atmosphere (10-50 km above the surface of the earth). It is characterised by the presence of large concentrations of the gas ozone O_{CO2} that constantly exchanges with oxygen through the action of sunlight and various compounds in the stratosphere. The body of ozone in the stratosphere is commonly referred to as the ozone layer. It is of crucial importance for the presence of life as we know it on earth, since it absorbs most of the ultraviolet (UV) radiations emitted from the sun, which can have very damaging effects, such as causing skin cancers.

It has been found that recent emissions of the man-made halo carbons (described in the paragraph on global warming) are contributing to the breakdown of ozone molecules in the stratosphere, thus favouring ozone degradation over ozone formation and reducing the extent and UV-absorbing capacity of the ozone layer. This may have serious direct effects on life on earth, especially on the plants and algae that are the primary producers and basis of all food chains. The possible indirect effects are at present hard to assess.

Acidification

Emissions of compounds that contribute to a lowering of the pH in some compartment of the environment are aggregated in this effect type. Acidification mainly occurs in areas where the underground strata are poor in limestone that may neutralise the incoming acids. The biological systems that are affected by acidification are lakes, forests and agriculture. The lowering of the pH in lake water or soil water dramatically changes the living conditions, resulting in massive death of the organisms normally living there (fish, trees, crop plants). Acidification also causes severe economic losses in urban areas, where corrosion of machinery and buildings may be substantially increased.

The contributors to acidification are compounds that are acids themselves (capable of donating protons in chemical reactions) or ones that cause liberations of protons as they experience conversions in a receiving body of the environment. Among the important sources of acidification is combustion of chlorine-containing fossil fuels, production of metal from sulphide ore, combustion of chlorine substances, e.g., in waste incineration.

Nutrient Enrichment

Nutrient enrichment is the process of enrichment of biotopes (waterbodies and soils) in the two main nutrients, nitrogen, (N), and phosphorus, (P), one of which is generally the nutrient factor limiting biological growth of the system. The effect type incorporates the ensuing changes in chemical and physical living conditions for the organisms and consecutive or cumulative disturbances of the ecosystems. The main focus is on the eutrophication of lakes and coastal waters. Here, the input of nutrients causes algal blooms that shadow out the plant life at the bottom of the waterbody, even in shallow waters, leading to a vast precipitation of dead algae. The degradation of this additional input of biological material uses up the oxygen dissolved in the water, causing anoxic conditions in the sediments and in the bottom strata. The effect on the fish life and the aquatic ecosystems may lead to dramatic reduction in catches of fish.

Nutrient enrichment of soil ecosystems may cause changes in the species composition of plant communities growing on soils poor in nutrients, thus threatening to extinguish certain plant species. Through percolation, the nutrient-enriched precipitation may reach the groundwater and cause unwanted concentrations of compounds, particularly of nitrogen, in this source of drinking water.

Contributing to nutrient enrichment are all the compounds containing nitrogen or phosphorus that may be released during their degradation in the environment. Important sources of *N* and *P* are agricultural fertilisers and emissions of untreated sewage from urban districts. Combustion processes may also be a significant source of nitrogen.

Photochemical Ozone Formation

Although neurotoxin in themselves, most hydrocarbons will rarely contribute to toxicity in the environment, because their atmospheric half-lives are too short for large concentrations to build up in the troposphere (0-10 km above the surface of the earth). The atmospheric degradation of the hydrocarbons takes place through reactions with nitrogen oxides and hydroxyl radicals present in the troposphere. The reaction schemes include photochemical reactions requiring sunlight, and they lead to the formation of aggressive gaseous compounds, the most important of which is ozone. While the presence of ozone in the higher strata of the atmosphere is indispensable for the existence of life, as we know it, it is highly noxious in the air that we breathe.

Episodes of high concentrations of these reactive gases may occur locally due to emissions of excessive amounts of hydrocarbons, e.g., from traffic during rush hours, but their extent may increase to a regional scale in periods where meteorological conditions prevent the vertical mixing of air within the troposphere. Such episodes are referred to as photo (or photo-chemical) smog.

Ozone is a very reactive molecule, and it will oxidise any reduced compound it encounters including most of bimolecular that are present in the troposphere. Plants in particular suffer from exposure to high ozone concentrations, as they are unable to move during episodes of photo smog, and they depend completely on the exchange of gases with the surrounding air to furnish carbon dioxide for photosynthesis.

Toxicity (chronic or acute)

This type of environmental effect includes contributions to toxicity (or poisoning) affecting any kind of living organism or ecosystem (human toxicity as well as environmental toxicity). The toxicity may be acute or chronic, and literally thousands of different chemical compounds may contribute, with toxic effects of many diverse types. Potential contributions to acute or chronic toxicity in the environment are much more difficult to quantify and to aggregate than contributions to any of the other effect types.

Since many different compounds contribute to this environmental effect type, most human activities that might be considered an object for an environmental assessment entail releases that will be toxic in the environment. Therefore, it is difficult to point out individual human activities as the main sources of environmental toxicity. Nevertheless, chemical industries of all types generally cause significant emissions of many different toxins affecting humans and the environment.

Though a quantitative approach to the aggregation or accumulation of toxicity is difficult, this item should under no circumstances be left out of an environmental assessment. Since toxic effects will generally occur on a local scale, the releases contributing to toxicity will often be the ones that are felt most strongly by people living in the surroundings of the project considered. They will, therefore, attract the most attention. Neglecting toxic effects in the classification may strongly reduce the general credibility of the environmental assessment.

Resource Consumption

Apart from reducing the environmental load of the product, a general goal of environmental management of products is to reduce the use of limited or non-renewable resources, so as to not impair the possibilities for future generations on planet earth. (Current practices should, as a general rule, be sustainable.) Therefore, a distinction should be made between renewable and non-renewable resources in the inventory. A renewable resource is defined as one that is being replenished naturally, whereas non-renewable resources are not regenerated at a rate that is significant in a historical perspective.

For historical reasons, use of resources in energy production is considered as a special kind of resource consumption with its own inventory. Also, an inventory of the consumption of energy resources must distinguish between renewable and non-renewable resources.

The bare classification into renewable and non-renewable resources is insufficient for assessment purposes. As a minimum, some further raking of the different kinds of non-renewable resources is necessary. For example, although both sand and copper ore are non-renewable resources, their use should not be considered as equally critical in an LCA. Unfortunately, there is no generally accepted scheme for classification of resources consumption, similar to the scheme adopted for environmental releases.

CASE 6.3: LIFE CYCLE ANALYSIS CASE STUDY: SUBSTITUTES FOR PVC

After a long, fierce political debate on the desirability and possibility of reducing the consumption of the plastic material PVC (Poly Vinyl Chloride), The Danish Agency for Environmental Protection commissioned a life cycle assessment of PVC and alternative materials to determine the environmental and health related consequences of substituting other materials for PVC in packaging materials, building materials and office and hospital articles.

The life cycle assessment (LCA) was performed as a screening involving all life cycle stage of the materials, from the extraction of raw materials to their disposal through landfilling, incineration or possible recycling. A more detailed study was performed on what seemed, from the screening, to be the key issues for the life cycle of the material.

This investigation differed from most other LCA studies in assigning equal importance to working environment/health/risk aspects and the external environment.

The system studied was specified by the prevalent conditions in Denmark concerning working environment and safety regulations. The disposal method was considered to be incineration with energy recovery for heating.

The following aspects were studied for each of the materials:

- consumption of resources (including energy),
- potential exposure in the working environment,
- potential effects in the working environment,
- potential exposure in the natural environment,
- potential effects in the natural environment, and
- risk of accidents.

The materials evaluated were:

- polyvinyl chloride, PVC,
- polyethylene/polypropylene, PE/PP,

- polystyrene, PS,
- polyurethane, PUR,
- polyethylene terephthalate, PET,
- vacuum-treated and surface-treated wood,
- paper/cardboard, and
- aluminium.

Several of the investigated materials were only considered as alternatives to PVS in some application. Some conclusions of the study are given below.

Polyvinyl Chloride

The worst problems related to environmental aspects seemed to be:

- emissions of dioxins and chlorinated organic compounds in the waste water from production and combustion gases from incineration of PVC;
- emission of hydrochloric acid and toxic heavy metals (particularly lead and cadmium) that may be added as stabilisers to PVC for some applications.

Those related to health aspects were:

- exposure to carcinogenic vinyl chloride monomer, which may occur during the production of the polymer and its manufacture into products;
- exposure to hydrochloric acid and to the carcinogen plasticiser, DEHP, occurring during manufacture of PVC products;
- risk of exposure to chlorine and carcinogenic vinyl chloride monomer in the event of accidents at PVC polymer production facilities.

Polyethylene/Polypropylene

Unlike PVC, these plastics do not contain chlorine, and this leads to substantial improvements in environmental and working environment performance. The energy consumption for production of PE and PP is similar to that required for PVC. The main problem with these materials is that they require addition of flame retardants from some applications, which may introduce environmental problems in the production, manufacture and disposal of the polymers.

The overall conclusion, however, is that the substitution of PVC with polyethylene or polypropylene, where possible, will give substantial improvements in external and working environmental performance.

Polystyrene

The production of polystyrene involves a risk of exposure to the reproduction-damaging substance, styrene. The polymerisation of styrene to polystyrene may involve a risk of explosion. The energy consumption for production of polystyrene is considerably higher than for PVC.

Polystyrene may be expanded into a rigid foam by blowing with gases. In some cases, the gases used are the highly ozone-degrading CFC or HCFC gases. If use of CFCs and HCFCs is avoided, the substitution of polystyrene for PVC will lead to environmental improvement, particularly at the disposal stage. The working environment aspects are unclear.

Polyethylene Terephthalate

PET entails somewhat higher energy consumption than PVC. The overall environmental performance is clearly better for PET than for PVC, both for working environment, external environment and probability and consequences of accidents.

Polyurethane

The production of PUR involves the use of isocyanides, exposure to which is extremely unpleasant in the working environment during production of the polymer and manufacture into products.

PUR may, similarly to polystyrene, be expanded by blowing with CFCs or HCFCs. If this application is disregarded, the potential effects of PUR in the external environment seem smaller than for PVC, particularly at the disposal stage (due to the absence of chlorine and toxic metals). For working environment conditions, however, there is some aggravation compared to PVC. Overall, the report does not recommend a substitution of PVC with PUR.

Vacuum-treated and Surface-treated Wood

In contrast to all the polymer materials, wood is a renewable resource and generally, little or no fossil energy is required for its production. The manufacture of vacuum-treated wood, e.g., into window frames, involves the exposure of works to carcinogenic wood dust, and the vacuum-treatment of the wood entails emissions of heavy metals into the external environment. Nevertheless, the overall conclusion qualifies vacuum-treated or surface-treated wood as an environmentally advantageous alternative to PVC.

Paper/Cardboard

Like wood, paper and cardboard are made from renewable resources. Their energy consumption is higher than for wood, but still lower than for PVC.

The environmental profile of paper and cardboard depends strongly on the process used. Often, the pulp is bleached using chlorine compounds, resulting in emission of chlorinated organic compounds (including dioxins) in the wastewater. Generally, pulp production involves emissions of large quantities of toxic and oxygen-consuming organic material in the wastewater that is worse than the loading from the PVC-production. The disposal of paper or cardboard after its use is substantially better from an environmental point of view, and overall, paper or cardboard is recommended over PVC for packaging purposes and similar applications.

Aluminium

Production of primary aluminium from bauxite is extremely energy-consuming and involves some serious working environment problems and hazards. Together these aspects disqualify the substitution of PVC with primary aluminium.

For secondary aluminium, i.e., aluminium recycled through collection and recasting, the situation is different since it may be recycled, and the loading of the original production of primary aluminium reduced correspondingly. The preference of PVC or aluminium thus depends on the efficiency of the recycling of aluminium.

CASE 6.4: LIFE CYCLE ANALYSIS: A CASE OF STEEL SECTOR IN INDIA

Since 1999, with an objective of optimum utilisation of resources, Life Cycle Assessment Study in the Steel Sector has been continued with joint funding of the Ministry and three identified steel plants, viz., Bhilai Steel Plant, SAIL; Jamshedpur Steel Plant of TISCO and Vizag Plant of Rashtriya Ispat Nigam Ltd. (RINL). The salient findings of the study are summarised herein.

Scope of the Study

- Cradle to Gate (RM mining to Crude Steel)
- Includes Inventory Analysis
- Impact assessment
- Improvement Analysis

Boundary of the Study

- Crude Steel at Factory Gate
- Quantification Unit-Kg of crude steel at factory gate
- Includes:
    - Steel Making
    - Production and Transportation of RM
    - Production and Transportation of Energy
    - Production and Transportation of Consumables

Data Collection and Validation

- Main Source of Data—Annual Statistics
- Coke Oven Gas, BF Gas and BOF Gas conversion from calories to joules by 4.1868 joules/cal
- Heat Content of 8 ata steam 3000 MJ/tonne and 18 ata 3260 MJ/tonne
- Fugitive Dust and drinking water not included
- Data Validation for FY1997 and FY 1998 by MECON
- Primary data validation by log sheet checks
- Secondary data validation—Monthly, quarterly, half yearly and annual Audits
- Energy Audits—Steam, Electricity, Fuel
- Emission Audits—Weekly stack and effluent analysis
- Performance Audits—Daily Production Target monitoring at department level and against techno-economic parameters at plant level
- Technology Audits

Results

Sl. No.	*Item*	*World Av. (Kg/Kg of steel)*	*Units Av. (Kg/Kg of steel)*
1.	Coal	0.536	0.989
2.	Dolomite	0.0291	0.158
3.	Limestone	0.0001	0.171
4.	Iron Ore	1.38	1.813
5.	Steel Scrap	0.142	0.110
6.	Water	8.1 (l/kg)	4.74 (l/kg)
7.	Energy	21.7 (MJ/kg)	29.77 (MJ/kg)

Emission Results

Sl. No.	*Item*	*World Av. (g/Kg of steel)*	*Units Av. (g/Kg of steel)*
1.	CO_2	1734	2924.65
2.	CO	26.4	63.09
3.	NO_x	1.57	1.39
4.	SO_x	1.48	1.70
5.	Particulates	1.57	1.275

Result Analysis

* Ash content of domestic coal high, so coal consumption high.

* Alumina in domestic ore high, so higher dolomite requirement.

* Iron ore lump used in cooling in BOF in India and nowhere else in the world, so higher iron ore consumption.

* Steel plants outside India mostly on shores, don't recycle water; hence high consumption.
* World over steel plants use BF gas top pressure to extract electricity, not in this unit.
* BF gas having higher carbon though lower calorific value compared to other gases used, so higher CO_2 emission.
* World over alternate fuels used in BF like oil, natural gas, pulverised coal, etc. BF gas generated per tonne is low.
* High CO emission in sinter making process.
* Low sulphur in Indian coals, so low SO_x emission.
* ODS use is low.

Impact Assessment

Global Warming effect (direct, 20 years)

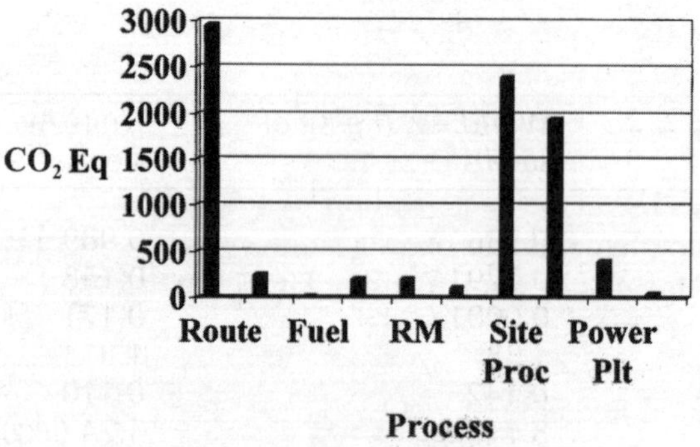

Air Acidification Effect

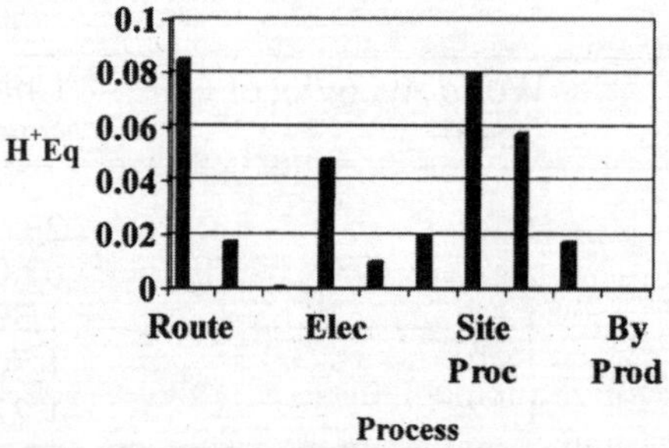

Improvement Planning

- Increase in proportion of imported coal in coal blend
- New sinter machine installation
- Installation of Top Gas Pressure Recovery to harness electricity
- Coal dust injection in blast furnace

UNIT 7

Environmental Management System Standards

STRUCTURE

OVERVIEW

In Unit 5, we discussed how to plan and conduct various types of environmental audits and then in Unit 6, we explained how to use life cycle assessment (LCA) as a tool for minimisation of environmental impacts associated with manufacturing processes and products. However, use of these tools assumes that an environmental manager is someone external to the industry, studying the system with an objective that the industry complies with the regulations and, on the basis of which, recommends measures for implementation. Establishing an organisational system conforming to environmental management system (EMS) standards, however, represent a paradigm shift away from the traditional approach based on command and control regulations, which set levels for pollutants. EMS standards focus on management practices, which help companies assess the level of their efficiency in meeting their environmental responsibilities. In other words, EMS is more of a foundation or template for environmental performance, allowing companies to put systems in place to address environmental concerns on many levels, and motivating, "to go beyond compliance". It also helps manage pollutants in unison as a whole, rather than in isolation.

Against this backdrop, in Unit 7, we will discuss the concept of EMS. We will also introduce you to EMS standards ISO 14000 and ISO 14001.

LEARNING OBJECTIVES

After completing this Unit, you should be able to:

- explain the benefits of establishing EMS;
- plan the process for establishing EMS;
- discuss ISO 14001;
- work as a co-ordinator for getting ISO certification for your organisation.

7.1 ENVIRONMENTAL MANAGEMENT SYSTEMS (EMS)

The growing public interest in environmental issues and concern for environmental quality has led to the emergence of strict pollution control regimes. This has brought about the development and implementation of various voluntary schemes and standards for environmental management and pollution control. Notable examples are BS 7750, the European Union's Eco-Management and Audit Scheme (EMAS) and the ISO 14000 series. In this context, putting in place systems to meet with these standards assume significance. An environmental management system (EMS) of an organisation is a system designed to:

- meet the regulatory and legislative system requirements;
- improve the control of the environmental impact;
- provide confidence to the customers that the products and services are manufactured with the aim of reducing the negative impact on the environment;
- suitably accommodate changing market trends and gain competitive edge;
- reduce the costs associated with environmental liabilities and insurance;
- gain public and media support.

EMS involves preparing a list of all environmental legislation, making a register of all environmental effects associated with the company's activities, setting targets (such as reducing waste production by a quantifiable amount), keeping appropriate records and undertaking regular reviews of the system. The whole system, in essence, forms the basis for sound environmental performance.

Most EMSs are built on the notational model illustrated in Fig. 7.1. This 'plan-do-check-act' model leads to continuous improvement based upon (*http://www.epa.gov/ems/info/index.htm*):

- planning, including identifying environmental aspects and establishing goals (plan);
- implementing, including training and operational controls (do);
- checking, including monitoring and corrective action (check);
- reviewing, including progress reviews and acting to incorporate required changes in the EMS (act).

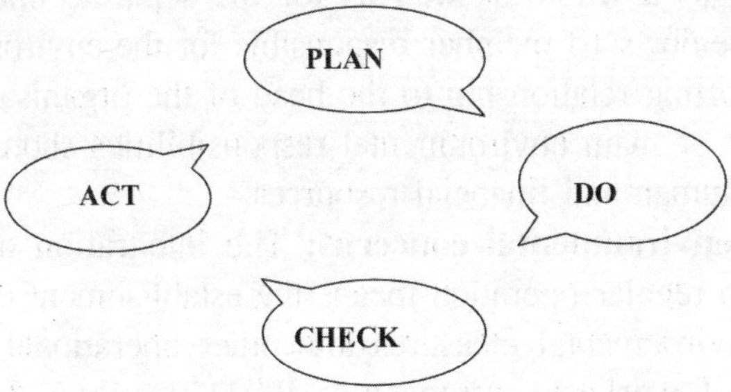

Figure 7.1 The Plan–Do–Check–Act Model

To reiterate, this model ensures that environmental matters are systematically identified, controlled and monitored. It also helps ensure that performance of the EMS is improved over time.

7.1.1 Core Elements of EMS

There is no single best approach to the development and implementation of EMS, since it depends on the nature, size and complexity of the activities, products and services within the organisation. All environmental management systems, however, have a number of core elements in common and these include the following:

- **Environmental policy:** An environmental policy is usually published as a written statement, expressing the commitment of the senior management to improving appropriate environmental performance. It is most often understood as a public statement of the intentions and principles of action for the organisation regarding the environment. The policy statement should define the broad environmental goals the organisation has decided to achieve. These are most clear if they are quantified, e.g., to reduce emissions of pollutants by 95% within five years, to provide sewerage and biological treatment of sludge for 60% of the population within three years, etc.
- **Environment action plan:** An environment programme or action plan describes the measures the organisation will take over the coming year(s). The action plan essentially translates the environmental policies of the organisation into objectives and targets and identifies the activities to achieve them, defines responsibilities and commits the necessary human and financial resources for implementation. This includes committing the necessary funds and staff to

meet each goal and providing for monitoring and co-ordination of the progress made towards achieving these separate goals and the overall policy goals that have to be fulfilled.

- **Environment related organisational structures:** The structures establish assignments, delegate authority and assign responsibility for actions. In the case of organisations with multiple sites or different activities, structures for the organisation as a whole as well as for the separate operating units are necessary. The senior staff member responsible for the environment, typically, has a direct reporting relationship to the head of the organisation. Individuals holding strategic or main environmental responsibilities should be adequately supported with human and financial resources.
- **Integration of environmental concerns:** The integration of environmental management into regular operation means the establishment of procedures for incorporating environmental measures into other operational aspects such as the protection of workers, purchasing, R&D, product development and acquisitions, marketing, finance, etc., in the case of companies and the safety, health and welfare of the community in the case of a local government. This encompasses the development of specific environmental procedures, usually detailed in operating manuals and other operating instructions describing measures and actions to take in the implementation of the environmental programme or action plan.

Figure 7.2 gives a schematic representation of these core elements.

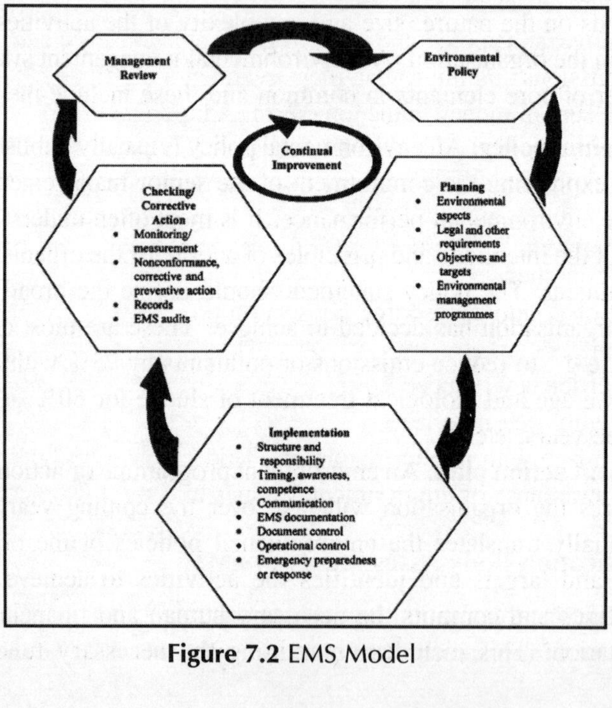

Figure 7.2 EMS Model

Due to the cyclic nature of the model, EMS is very dynamic in that a change or revision in any of the core elements of the system will have a chain/cascading effect on other elements.

Note that while there are several EMS models available, the model presented in Fig. 7.2 uses the standards set by ISO 14001 as a starting point for describing EMS elements. This is so, mainly, due to the fact that:

- ISO 14001 is a widely accepted international standard for EMS that focusses on continual improvement.
- Companies may be asked to demonstrate conformance with ISO 14001 as a condition of doing business in some markets.

7.1.2 Benefits of EMS

A number of benefits are associated with the adoption of EMS. For example, insurance companies are beginning to exclude pollution incidents from their insurance cover, and when they are included, the insurance companies place a surcharge on the policies. In order to improve or reduce any increases in the insurance premiums, companies can carry out an environmental audit and implement an EMS.

From a managerial point of view, an EMS enables a company to develop and maintain a well-organised management structure that ensures compliance with environmental legislation. By keeping the working place cleaner and safer, and the production more efficient, staff interest can be aroused and morale improved. Another incentive is the potential financial gains, which can be achieved. Reducing wastes and emissions and improving their treatment helps many companies to save a considerable sum of money in the long term. Adopting an EMS will also facilitate trade, as major trading blocs are more likely to accept members with a certain level of environmental performance. In other words, the benefits of an EMS include:

- Improved environmental performance.
- Enhanced compliance.
- Prevention of pollution/resource conservation.
- New customers/markets.
- Increased efficiency/reduced costs.
- Enhanced employee morale.
- Enhanced image with public, regulators, lenders, investors.
- Employee awareness of environmental issues and responsibilities.

7.1.3 Certification Body Assessments of EMS

To be certified, an organisation has to demonstrate that the EMS functions and the various control mechanisms are properly operational. In practice, this means that the EMS has been operational for a minimum of three months, the internal audit system is fully operational and that one management review has been conducted (*Tech Monitor,* 1995).

Based on the following conditions, certificates are issued:

- Non-conformities identified have been addressed and eliminated.
- The certification body has justified confidence that all provisions in the EMS standard have been met, and, in particular, that provision for compliance with the organisation's policy and objectives is effective.
- The principle of continuous improvement of environmental performance has been stated within an environmental programme and is being adhered to play in the EMS.
- All key staff (i.e., those involved in managing significant effects) have undergone a training need analysis and have received training.

Regulatory Compliance

An organisation with a certified EMS has a system that should achieve continuing compliance with regulatory requirements applicable to the environmental aspects of its activities, products and services. The certification body confirms that a system designed to achieve the necessary compliance is operating effectively. In situations where authorities issue licenses or permits, it will often be a requirement that the organisation notifies the authority of any breaches. The certification body should ensure that the organisation's EMS records any infringement of regulations and that the appropriate corrective action has been taken.

An organisation with a certified EMS is responsible for continuing compliance with all the regulatory requirements and, therefore, must maintain a record of incidents of non-compliance and the remedial action.

Surveillance Audits

At each surveillance, the certification body should pay attention to the following:

- the effectiveness of the EMS with regard to achieving the objectives of the organisation's environmental policy;
- an interview with management responsible for the EMS;
- the functioning of procedures for notifying the authorities of any breaches;
- progress of planned activities aimed at continuous improvement of environmental performance, where applicable;
- follow up conclusions resulting from internal audits;
- action taken on non-conformities identified during the previous audit.

The certification body should be able to adapt its surveillance programme to the environmental issues related to the activities of the organisation and justify this arrangement. The surveillance audit of the certification body should be agreed with the organisation, taking into account the internal audit programme and the reliability that can be attributed to this work.

7.1.4 Documentation for EMS

While data relates to fact, information refers to a series of data analysed to provide a decision. Information is an answer to a question. All information comes from data, but not all data come from information (*Tech Monitor,* 1995). In other words, executives making decisions often receive excessive amounts of data, but they do not always get the information they need, presented clearly and adequately for use in decision-making.

Decisions are usually related to solving problems. In order to solve a problem, one needs information to decide. An EMS, therefore, requires a proper documentation system in order to collect, analyse, register and retrieve information.

The documentation should describe the EMS and make clear the relationship to any other management system in operation within the organisation, or as having an influence on the operations of the entity, subject to certification. It is acceptable to combine the documentation for environmental and other management systems (such as for quality of health and safety), as long as the EMS can be clearly identified, together with the appropriate interfaces between different EMS documentations and the order of precedence.

Environmental effects may have been considered within safety reviews, hazard and operability studies and risk/consequence analysis, etc. In such cases, the EMS documentation should refer to the critical areas where significant effects are covered by other management systems. The EMS documentation should be available to all appropriate staff and should be kept up-to-date.

Environmental Management Manual

A key EMS document is a company's environment manual. This is a document that establishes the general policy of a company on this issue. It usually contains an environmental policy and a clear statement on the person who is in charge of verification of activities in the organisational structure pertaining to the maintenance of the environmental system. It lays down how each requirement of the policy will be handled in the organisation and includes a list of all procedures.

The manual also contains a description of the system, which was or is being implemented. There are considerable benefits in preparing a manual, which:

- clarifies responsibilities;
- is useful for audit purposes;
- can be highly effective for training of new employees;
- makes easier the process of obtaining approval of licenses from protection agencies;
- is useful for marketing purposes.

The highest executive in the company must sign the document. It should contain the vision of the organisation about environment and its commitment to continuous improvement. The document should detail targets to be reached and explain how they will be reached, with what resources and under whose responsibility. The company must also realise that it is necessary to measure results in order to evaluate how correctly the policy is being applied. The company should, therefore, have a programme for achieving its objectives and targets.

To prepare the manual, it is necessary first to identify all mandatory regulations, standards and codes to be complied with. In addition, the company needs to identify other requirements, such as headquarters' policies, top management's strategic vision, market demands, etc.

The discussions that we have had so far provide you with a basic knowledge of EMS. Now, we will next discuss the importance of ISO 14000 and its supporting systems. Before we do so, let us first do Learning Activity 7.1.

LEARNING ACTIVITY 7.1

Can you predict the advantages of EMS over an environmental audit as an environmental management tool?

Note:

a) Write your answer in the space given below.

b) Check your answer with the one given at the end of this book.

7.2 EMS STANDARD: ISO 14000 SERIES

Companies throughout the world have begun to understand the requirements of an environmental management system (EMS) as well as the benefits such a system can provide. However, given the nature of company dynamics, changes to business priorities, personnel and work patterns, it is imperative to introduce a proper management system which maintains the focus on environmental improvement through the changing landscape of corporate activity. It is against this background that ISO 14000 Series are being developed and implemented. In Subsections 7.2.1 to 7.2.3, we will discuss the various aspects of ISO 14000 series.

7.2.1 Evolution

The International Organisation for Standardisation (ISO) was formed in 1947 and has since become the premier international standards organisation. Since its inception, ISO's mission has been to promote worldwide standardisation in order to facilitate international commerce. The organisation began by developing international agreements and in 1951 published its first standard—Standard Reference Temperature for Industrial Length Measurement. Since that time, ISO has developed more than 9000 standards for a variety of subjects ranging from screw threads and fasteners to high-tech clean rooms. ISO also developed the widely used quality management system (QMS) standards, i.e., the ISO 9000 series. The ISO standards are published for voluntary acceptance, but they are often incorporated into national standards of individual countries.

The membership of ISO includes over 100 countries. Each member country is represented by one standards organisation. For example, the American National Standards Institute (ANSI) represent the U.S and Great Britain by the British Standards Institute (BSI).

ISO considers the following three key principles in developing international standards:

(i) **Consensus:** The views of all interested parties are taken into account, including manufacturers, vendors, consumer groups, testing laboratories, governments, engineering professionals and research organisations.

(ii) **Industry-wide applicability:** The goal is to draft standards that satisfy industries and customers worldwide (ISO has no authority to impose its standards on any government or organisation).

(iii) **Voluntary nature:** All of the international standards developed are voluntary. Thus, their acceptance by industry is market-driven and based on voluntary involvement of all interests in the marketplace.

A review of historical developments reveals that standardisation on a worldwide basis was accelerated with quality and occurred generally independent of environmental management. Attempts to standardise quality requirements were made by many organisations. However, it wasn't until the 1990's that a significant level of agreement was reached.

In 1990, Business Charter for Sustainable Development (BCSD), an organisation of fifty business leaders with interest in environment and development issues, was created for environment protection with the premise that economic development can take place only in a healthy environment. Partly in response to the proliferation of differing environmental standards, such as EMAS worldwide, the ISO formed a Strategic Action Group on the Environment (SAGE) in 1991.The purpose of SAGE was to investigate the usefulness of drafting international standards for environmental management. SAGE focussed its attention on the following three areas:

(i) Promoting a common worldwide approach to environmental management in business and industry.
(ii) Increasing the ability of incentives for organisations to measure and attain improvements in environmental performance.
(iii) Facilitating world trade and removing potential environmental trade barriers.

As a result of the findings of SAGE, the ISO formed the Technical Committee 207 (TC 207) in 1993. TC 207 became responsible for drafting the ISO 14000 series of standards. At early meetings of TC 207, more than thirty countries and 200 representatives expressed a desire to develop new EMS standards. TC 207 itself had members representing some sixty-nine countries and was divided into six sub-committees. These members included representatives from various industries, standards organisations, governments, environmental organisations, and other interest groups.

For some, the motivation for the development of these standards was due to the fear that the increased number of inconsistent national and regional EMS standards would create trade barriers. There was also the concern that the EMAS programme already in practice in Europe would influence the ISO 14000 standards to make them comparable to EMAS. This concern was in part fueled by the fact that the ISO 14000 standards were being developed in conjunction with CEN, i.e., the European standards-setting body. In fact, the ISO set the standards drafting timetable at 30 months, in part because CEN had agreed to accept ISO 14000, if they were finalised quickly enough. There were also fears that the standards were a reflection of European and not American technology. Nevertheless, the standards were developed with each participating member having equal say in the process.

A sub-committee of TC 207 prepared the draft EMS Standard, based on BS 7750. It was presented at the Earth Summit in Rio de Janeiro, and the draft of the standard was agreed in Oslo (June 1995).

7.2.2 Principles and Structure

The aim of the ISO 14000 series of standards is to help organisations implement and improve their EMS.

Some of the principles governing ISO 14000 series are (*Tech Monitor,* 1995):

(i) Understand all activities and processes being undertaken by the organisation.
(ii) Identify potential aspects associated with the activities of the unit at all stages and determine their impact on the environment.
(iii) Determine processes/procedures/operation steps that can be controlled to eliminate or minimise the likelihood of an occurrence of the adverse impact.
(iv) Identify the regulatory requirements relating to them and establish target level and tolerances, which must be met to ensure that operations affecting the environment are under control.
(v) Establish a monitoring mechanism to ensure control of these aspects.
(vi) Establish corrective actions to be taken when monitoring indicates that a particular aspect is not under control.
(vii) Establish a system of emergency preparation and for meeting such exigencies.
(viii) Establish procedures for verification to confirm that the environmental management system is working effectively in compliance with regulations and recording continuous improvement.
(ix) Establish documentation concerning all procedures and records appropriate to these principles and their application.

Given the importance of the stakes involved and the generic nature of the international requirements covered, high expectations have been placed on the ISO 14000 series standards (*Tech Monitor,* 1995). The standards can be classified into the following two categories based on their focus:

(i) **Organisation or process standards:** These include environmental management system (EMS), environmental auditing (EA) and environmental performance evaluation (EPE).
(ii) **Product-oriented standards:** These include life-cycle assessment (LCA), environmental labelling (EL) and environmental aspects in product standards (EAPS).

LEARNING ACTIVITY 7.2

List the basic principles of ISO standards.

Note:

a) Write your answer in the space given below.

b) Check your answer with the one given at the end of this book.

The ISO 14000 series standards are of two types, and these are:

(i) **Normative standards:** These indicate the requirements that must be met and can be audited for certification.

(ii) **Informative standards:** These provide guidance and the requirements need not be audited for certification.

In the 14000 series of standards, ISO 14001 (i.e., Environmental Management Systems—specification with guidance for use) is the only normative standard and all other standards are informative standards intended to support the implementation of EMS.

As regards the structure of the Series, ISO 14001, i.e., the requirements for EMS, forms the nucleus, and in the first orbit is ISO 14004, i.e., the guideline standard. (Note that, we will discuss ISO 14001 standards in detail in Section 7.3.) The other standards in the series are supporting systems as shown in Fig. 7.3.

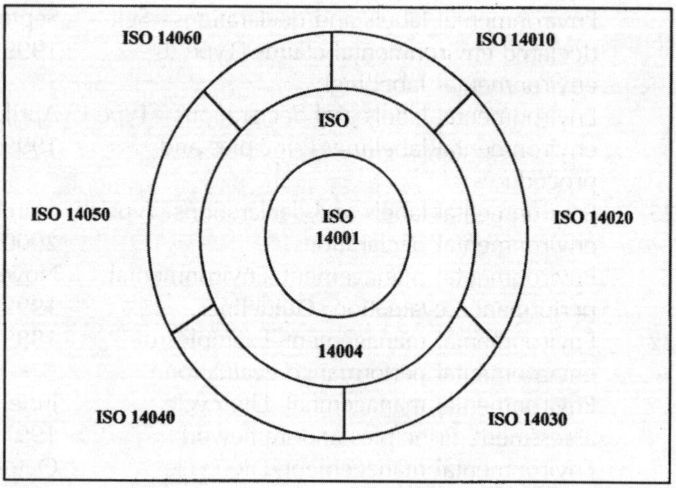

Figure 7.3 Structure of ISO Series and Inter-relationships

7.2.3 Supporting Systems

The supporting systems of ISO 14000 consist of ISO 14010, ISO 14020, ISO 14030, ISO 14040, ISO 14050 and ISO 14060. (*Tech Monitor,* Vol. 15, No. 5, Sept-Oct 1998). In what follows in this Subsection, we will discuss some of these supporting systems.

Consider Table 7.1, which gives a list of ISO 14000 Series standards with their respective publication dates.

Table 7.1 ISO 14000 Series Standards

ISO No.	*Title of International Standard/Guideline/ Technical Report*	*Publication Date*
ISO 14001:1996	Environmental management systems–Specification with guidance for use.	September 1996
ISO 14004:1996	Environmental management systems–General guidelines on principles, systems and supporting techniques.	September 1996
ISO 14001:2004	Revision of ISO 14001:1996.	November 2004
ISO 14015	Environmental management–Environmental assessment of sites and organisations (EASO).	November 2001
ISO 19011	Guidelines for quality and/or environmental management systems auditing.	July 2001
ISO 14020 2nd Edition	Environmental labels and declarations–General principles.	September 2000
ISO 14020: 1998/DAM1	Draft amendment 1 to ISO 14020:1998.	December 1998

ISO 14021	Environmental labels and declarations—Self—declared environmental claims (Type II environmental labelling).	September 1999
ISO 14024	Environmental labels and declarations—Type I environmental labelling: Principles and procedures.	April 1999
ISO/TR 14025	Environmental labels and declarations–Type III environmental declarations.	March 2000
ISO 14031	Environmental management–Environmental performance evaluation: Guidelines.	November 1999
ISO/TR 14032	Environmental management–Examples of environmental performance evaluation.	1999
ISO 14040	Environmental management–Life cycle assessment: Principles and framework.	June 1997
ISO 14041	Environmental management–Life cycle assessment: Goal and scope definition and inventory analysis.	October 1998
ISO 14042	Environmental management–Life cycle assessment: Life cycle impact assessment.	March 2000
ISO 14043	Environmental management–Life cycle assessment: Life cycle interpretation.	March 2000
ISO/WD TR 14047	Environmental management–Life cycle assessment: Examples of application of ISO 14042.	1999
ISO/CD TR 14048	Environmental management–Life cycle assessment: Life cycle assessment data documentation format.	1999
ISO/TR 14049	Environmental management–Life cycle assessment: Examples of application of ISO 14041 to goal and scope definition and inventory analysis.	March 2000
ISO 14050	Environmental management–Vocabulary.	May 1998
ISO 14050: 1998/DAM1	Draft amendment 1 to ISO 14050:1998.	December 1999
ISO/TR 14061	Information to assist forestry organisations in the use of Environmental Management System standards ISO 14001 and ISO 14004.	December 1998
ISO/AWI 14062	Guidelines for integrating environmental aspects into product development.	To be determined
ISO Guide 64	Guide for the inclusion of environmental aspects in product standards.	March 1997

Environmental Auditing (ISO 14010, ISO 14011 and ISO 14012)

An environmental audit is defined as the systematic documented verification process of objectively obtaining and evaluating audit evidence to determine whether specified environmental activities, processes, conditions, management systems, or information about these matters conform with audit criteria, and communicating the results of this process to the clients. The guiding principles of audits include:

- basing the audit on defined objectives and drawing inferences based on analysis, interpretation and documentation of appropriate information;
- utilising an audit team that is independent of the activities they audit and utilising an auditor who meets the specifics of qualification criteria;
- exercising of due professional care by the auditor to maintain confidentiality and adequate quality assurance;
- using appropriate procedures for an objective audit;
- developing audit criteria, evidence and findings;
- ensuring that the process provides a desired level of confidence in the reliability of the audit findings and conclusions;
- providing an adequate report of findings.

Note that ISO 19011 provides guidance on general principles for conducting audits, and elaborates on the framework for the conduct of audit in order to ascertain if the organisation is doing what it says it will do and whether the EMS conforms to ISO 14001. It also provides guidance on auditor qualification criteria, including education, training, work experience, personal attributes and skills, maintenance of competence and due diligence.

Environmental Labelling Standards (ISO 14020)

ISO 14020 is based on a voluntary environmental labelling standard that requires a third party verification and is designed to reduce burdens arising from diverse multiple labelling. Existing eco-labelling schemes, based on government initiatives, aim at influencing customer decisions to select environmentally friendly products, but they lack the application of uniform criteria. These are, therefore, difficult to comply with and are potential trade barriers. The coverage of ISO 14020 is broad and includes goods or services for consumer, commercial and industrial purposes.

The guiding principles and practices for third-party environmental labelling programmes include the following:

- Standard and criteria applicable for environmental labels must be developed through a consensus process, and the programme must be voluntary.
- A product can be considered for environmental regulations of the country in which it is manufactured and marketed.
- Environmental labelling programmers should be selective and should distinguish leading product alternatives.
- The product criteria developed by the certification agency should be periodically reviewed to account for new developments and technologies.

- The process should be transparent with regard to criteria, certification and award procedures. The criteria should be revised periodically and clearly demonstrate that the funding sources for the programme do not create conflicts of interest or undue influence.
- The labelling programme should use scientific and reproducible methodologies to assess the environmental impacts of products.
- Labelling programmes should be accessible, objective and affordable.

The requirements for awarding a label are divided into general rules that apply to all products and applications, and specific product criteria that set requirements for each product category. The specific product criteria are the only criteria that may be considered as a basis for awarding the label. The certification agency awards the label when satisfied that the applicant has complied with the specific product criteria for the category. It maintains a publicly available list of products currently licensed to carry the label.

After the label has been awarded, it is the responsibility of the certification agency to take all necessary steps to ensure ongoing compliance with the product criteria. The certification agency will require the licensee to take corrective action, if monitoring indicates that compliance is not being maintained.

Environmental Performance Standards (ISO 14030)

The environmental performance evaluation (EPE) is an important ongoing internal management process. It uses environmental indicators to compare an organisation's past and present environmental performance with its environmental objectives, targets or other intended levels of environmental performance. This process helps management to measure, analyse, assess, report and communicate an organisation's environmental performance over time and to determine necessary actions.

The environmental performance evaluation standard is based on Edward Deming's PDCA cycle, i.e., Plan (P), Do (D), Check (C) and Act (A). Planning an EPE involves considering management priorities and selecting environmental indicators before evaluating environmental performance (D), and reviewing and improving EPE (C). The EPE process involves actions taken (A) to improve the systems, operations, processes and environmental performance. Note that we discussed the PDCA cycle in Section 7.1 (see Fig. 7.1).

EPE standards provide guidelines on identification and selection of environmental indicators. There are two types of environmental indicators, and these are:

(i) **Environmental performance indicators (EPI):** These include the following:

- the people, practices and producers at all levels of the organisation;
- the design, operation and maintenance of, supply to, and delivery from, an organisation's facilities and equipment;
- the materials, energy, products, services, waste and emissions related to the organisation's operations and activities.

(ii) **Environmental condition indicators (ECI):** These are used to describe the conditions of the environment in relation to the organisation.

Life Cycle Assessment (ISO 14040)

ISO 14040 provides guidelines for incorporating life cycle assessment (LCA) into environmental management programmes (recall that in Unit 6 we discussed LCA in detail). An LCA is defined as a compilation and evaluation of the inputs, outputs and the potential environmental impacts of a product system throughout its life cycle. It is, thus, a systematic set of procedures for compiling and examining the inputs and outputs of materials and energy and the associated environmental impacts directly attributable to the functioning of a product or service system throughout its life cycle from the acquisition of raw materials through final disposal. An LCA is done in order to get the whole picture of the environmental impacts throughout the lifetime of products and services. In other words, an LCA provides significantly more information than does evaluating the impact from the manufacturing process alone. It also provides a systematic way to evaluate the costs and benefits associated with products or service changes at various points in the life cycle.

ISO 14040 covers the following three phases of LCA:

- establishing the goals and the scope of the assessment;
- conducting inventory analysis.
- conducting impact and improvement assessment.

The goals of the LCA study should include the reasons for carrying out the study, the intended applications, the intended audience, the initial data, quality objectives and the type of critical review that will be conducted for the LCA. The scope should include background information for the product or service being evaluated, boundaries of the study, method of impact assessment, data requirements, assumptions and limitations of the study.

Environmental Terms and Definitions (ISO 14050)

The successful operation of the environmental management system requires a correct understanding of the meaning of terms used by all stakeholders in a similar manner. Put differently, communication is important in implementation and in the operation of the environmental management system. This communication will be most effective, if there is a common understanding of the terms used. Many environmental terms and definitions are the result of a recently developed process. The gradual evolution of these environmental concepts invariably means that environmental terminology will continue to develop.

Environmental Aspect in Product Standard (ISO 14060)

The environmental impacts of products are receiving more attention worldwide. They are, therefore, being addressed in the product standards prepared by the international, regional and national standard bodies. ISO 14060 provides general guidelines that should be taken into account when developing standards to reduce environmental effects associated without sacrificing the intended performance of the product or service. While recognising the complexity of identifying and establishing environmental effects from products and services throughout their life cycles, ISO 14060 provides that environmental effects should be balanced against factors that include product function, performance, safety and health, cost, marketability and quality. The guide also recognises that environmental provisions in product standards must be reviewed and changed to reflect innovation and technology, but not so frequently that innovation, productivity and environmental improvements are jeopardised. The provisions of a product standard should be no more stringent than necessary to avoid excessive or inefficient material or energy use.

The environmental provisions in product standards can incorporate considerations including material and energy inputs; wastes and emissions; impacts from transportation, packaging, distribution and use; reuse and recycling potential and product disposal and associated wastes. The impact can be assessed using LCA, risk assessment or other appropriate methods, provided that the method is appropriate to the product or service and the data is available for conducting the assessment.

Product standards use the following three strategies to protect the environment:

(i) **Resource conservation:** This focusses on preventing depletion of resources. Renewable resources can be replenished, if managed carefully. Non-renewable resources present different challenges. Recycling and waste minimisation are the two means that can be used to preserve these resources and to extend them as much as possible.

(ii) **Pollution prevention:** This utilises strategies such as source reduction and material substitution to reduce emissions into the environment. Pollution prevention can result in numerous financial benefits, including lower cleanup and disposal costs, reduced costs for pollution control equipment and its operation, insurance savings and many others.

(iii) **Design for environmental impacts:** This is established by conducting failure mode and effect analysis (FMEA).

LEARNING ACTIVITY 7.3

State the difference between ISO 14001 and other standards in ISO 14000 series of standards.

Note:

a) Write your answer in the space given below.

b) Check your answer with the one given at the end of this book.

ISO 14001 is what is implemented in companies. We will discus this in Subsections 7.3.1 to 7.3.3.

7.2.4 EMS Specification Standards: ISO 14001

This standard is the core of the ISO 14000 series. ISO 14001 presents the elements of an EMS that organisations are required to conform with, if they seek third-party certification. ISO 14001 was first published in 1996 and revised in 2004. As mentioned previously, ISO 14001 is the only standard in the ISO 14000 series intended to be used for third-party certification. All the other standards in the series are intended as guidance only. In development of your EMS, ISO 14001 will be your most important document—you will refer to it frequently!

The standard consists of the EMS specification in 17 general requirements (referred to as clauses) and three annexes.

Clauses

Since the standard has been written to be applicable to all types of organisations and to accommodate diverse geographical, cultural and social conditions, the requirements or clauses, are generalised and are neither specific nor prescriptive. That is to say, they are descriptive outlining the desired outcome of the system, but do not prescribe specific approaches that an organisation must implement to get there.

Annexes

Annexe A provides additional information on the requirements of ISO 14001 and is intended to help avoid any misinterpretation of the actual specification. It is expected that in future revisions to ISO 14001, the information in Annexe A may be either incorporated into the clauses or moved into ISO 14004. Annexe B identifies links and broad technical correspondences between ISO 14001 and ISO 9001 in a tabular form. The objective of this annex is to demonstrate how an EMS could be combined with a quality management system. Annexe C simply presents a bibliography of documents from both the ISO 9000 and ISO 14000 series.

Note that these annexes are solely for information and not for third-party certification. ISO 14001 is a management system standard. It is not a performance or product standard, although the framers developed it with the idea that the result of its implementation should give better industry environmental performance. The standard represents a shift towards holistic proactive management and total employee involvement. ISO 14001 urges employees to define their environmental roles from the bottom up and requires the support of senior management. It is a comprehensive framework that contains core elements for managing an organisation's processes and activities to identify significant environmental aspects the organisation can control, and over which it can be expected to have an influence. Any organisation or facility or organisation of any size anywhere in the world can use the standard.

An EMS conforming to ISO 14001 contains the following elements:

- An environmental policy supported by senior management.
- Identification of environmental aspects and significant impacts.
- Identification of legal and other requirements.
- Environmental goals, objectives and targets that support the policy.
- An environmental management programme.
- Definition of roles, responsibilities and authorities.
- Training and awareness procedures.
- Process for communication of the EMS to all interested parties.

- Document and operational control procedures.
- Procedures for emergency response.
- Procedures for monitoring and measuring operations that can have a significant impact on the environment.
- Procedures to correct non-conformance.
- Record management procedures.
- A programme for auditing and corrective action.
- Procedures for management review.

Note that while ISO 14001 is more likely to be used in manufacturing or processing industries, it can also be applied to services such as construction, architecture, health care and engineering. This standard essentially requires an organisation to state what it does in environmental management and do what it states. ISO 14001 is neither a product standard nor an environmental performance standard. It does not require an organisation to establish or disclose performance or audit results and does not require certification.

Let us now discuss the guidelines for planning and implementation of ISO 14001.

7.3 IMPLEMENTATION OF EMS CONFORMING TO ISO 14001

The key to a successful ISO 14001 EMS is having documented procedures that are implemented and maintained in such a way that successful achievement of environmental goals is commensurate with the nature and scale of the activities.

There are three components to an EMS and these are:

(i) **A written programme:** The written programme requires an organisation to be committed to producing a quality product with the lowest possible environmental impact and sets forth the procedures to be followed to achieve this goal.

(ii) **Education and training:** The implementation will not be effective, unless all employees have access to and understand the EMS, and this is achieved through education and training.

(iii) **Knowledge of relevant local and central environmental regulations:** The EMS must incorporate the applicable environmental legal requirements.

In addition, the EMS must include appropriate monitoring and review to ensure effective functioning of the EMS and to identify and implement corrective measures in a timely manner.

The various steps involved in the establishment of an EMS are as follows:

(i) **Obtain management commitment:** Senior management can demonstrate its commitment by playing an active, visible role in the EMS implementation process, providing funding and allocating resources, and promoting employee awareness and motivation. An EMS should be viewed as a tool to achieve continuous environmental improvement, rather than daily fire-fighting just to meet the mandatory requirements.

(ii) **Review current environmental programmes:** Once the management commitment is secured, the next step is to conduct and document a preliminary review of the organisation's current environmental programmes and management systems. The process includes reviewing existing environmental management policies, operating procedures and training programmes as well as the methods for identifying regulatory requirements.

(iii) **Conduct gap analysis:** Organisations need to evaluate as to what extent the existing procedures conform to the requirements of ISO 14001. Such a gap analysis identifies the actions necessary to build an EMS. To do this, organisations would need to compare their existing procedures to the following main stages of ISO 14001:

- Environmental policy.
- Planning.
- Implementation and operation.
- Checking and corrective action.
- Management review.
- Continuous improvement.

(iv) **Identify environmental aspects:** All parts of the planning phase contribute to identifying the significant environmental aspects of the operations. An environmental aspect refers to any element of an organisation's activities, products and services that can interact with the environment. ISO 14001 also refers to impacts. Impacts are the actual or potential changes to the environment resulting from any of the environmental aspects. Thus, the relationship between aspects and impacts is one of cause and effect. One of the best ways to identify the environmental aspects of the operation is by developing a process map or flowchart. This exercise involves mapping every step of the process and the inputs and outputs associated with each step. Developing a process map is best accomplished as a team effort, and therefore, representatives of a variety of departments and at diverse levels should be included to produce the most accurate description of the inputs and outputs. Note that an organisation's EMS is a team effort.

(v) **Identify significant aspects:** An organisation can utilise the expertise of staff to determine the significant environmental aspects of its operations.

This will help in focussing on those aspects over which the organisation has some control or influence. Determining which aspects are significant is based on the judgement of severity of the impact caused by the aspect and the frequency of its occurrence, and therefore, each aspect needs to be scrutinised. It requires many iterations, before the team arrives at a final list of significant environmental aspects.

(vi) **Decide the environmental policy:** Having identified the significant environmental aspects and established the means and norms to improve environmental performance in the form of objectives and targets, organisations need to write their broad intentions concerning the environment. This will form or evolve into an appropriate policy, meeting the requirements of the standards. This document must include the organisation's commitments towards complying with legal requirements, pollution prevention and continuous improvement. Note that policy framing may not be a one-time affair. It is quite likely that organisations are required to revisit the objectives formulated and targets identified commensurating with the outcomes of the initial implementation process. This iterative process helps in firming up the policy statements.

(vii) **Identify environmental management programmes:** An environmental management programme (EMP) may be considered a project or a plan of action to achieve objectives and targets. At this stage, the methodology of realising the objectives, the steps involved, the persons responsible for implementing these steps, the time frame of implementation, the means required, etc., should be noted.

(viii) **Identify training needs and personnel whose work affects the environment:** It is essential to identify the training needs of the personnel by comparing the present competence (based on the educational qualifications, past work experience, etc.) with the required competence. Organisations are expected to have training needs, training plans, training modules and training records, and must check the effectiveness of the training imparted. The training must make the personnel aware of the importance of conformance with environmental policy, environmental impacts of their work activity and the consequences of departing from the specified operating procedures.

(ix) **Establish communication system:** Proper communication channels are imperative to impress on the employees, as well as external interested parties, the reasons for adopting EMS.

(x) **Create proper documentation:** The EMS documentation describes the interaction among the core elements of the system and provides the key information necessary to understand the environmental management systems in place at the organisation. This is of immense use to the new employees and third-party auditors.

(xi) **Conduct EMS audits:** Organisations are required to conduct EMS audits to verify the EMS implementation. But, EMS audits conducted merely to meet the requirements of the ISO standard will not be of much use in the long term. For sustained benefits, organisations must invest in EMS audit programmes by giving employees proper training, providing appropriate time for audit, etc. Importantly, senior management should demonstrate its commitment to audits and provide visible support to the auditors. This will encourage the auditors to point out the weaknesses in the system without any hesitation. As inept audits—as a result of doctored auditing—are not uncommon, the support of senior management is very critical in carrying out and documenting EMS audits.

(xii) **Conduct management review:** Management must periodically review the EMS to evaluate its suitability and effectiveness.

(xiii) **Perform a final gap analysis:** Final gap analyses help organisations determine whether they are ready for a registration audit. In other words, the results of the final gap analyses help organisations decide whether or not to apply for certification by outside auditors.

Registration

Organisations are registered by outside auditors—also known as registrars—meeting the ISO 14001 standard. The registration can be for a specific site, several sites or the entire organisation. The registration process involves the following steps:

(i) Establishing the scope of the registration. This can be done internally or with the help of outside auditors.

(ii) Applying for registration. This is a written application to the registration body.

(iii) Reviewing documents about the organisation's EMS and submitting them to the registration body. Outside auditors perform full assessment.

(iv) Getting the grant of registration by the certification body based on the auditor's findings.

(v) Receiving either approval (conditional or provisional) or disapproval.

(vi) Receiving a certificate as a proof of registration and getting listed in a certification register.

(vii) Maintaining the registration through ongoing activities including monitoring and measuring the EMS, investigating and handling non-conformances, implementing corrective and preventive actions, maintaining environmental records and establishing and maintaining the ISO 14000 audit programme.

The time required to implement ISO 14001 EMS depends on the current status. That is to say, if an organisation's environmental programmes are effective and efficient and they are willing to commit resources, the EMS development will be fast. If all the factors required for the EMS development are in place, it is possible to complete the entire process till certification in eight to ten months. (Note that it may take the equivalent of one fourth of an employee's time for preliminary planning and exploring issues. The final four to five months will be required for checking and improving the established system. The longest and most time-consuming activity in the entire process will be training.)

Before you read any further, you must note that:

- ISO does not assess or audit environmental management systems to confirm that they conform to ISO 14001.
- ISO does not issue ISO 14000 certificates.
- ISO does not carry out ISO 14000 certificates (these are issued by certification (registration) bodies independently of ISO).
- The ISO logo is a registered trademark and ISO does not authorise its use in connection with ISO 14000 certification (registration).
- ISO 14001 is not a label signifying a *green* or *environmentally friendly* product. (No product label, advertisement or other promotional material should give the impression that a product is "ISO 14001-certified" or "ISO 14001-registered". Advertisements, which imply a quality ranking, i.e., ISO 9001, ISO 9002 and ISO 9003, are misleading.)

LEARNING ACTIVITY 7.4

Explain the terms *environmental* policy, *objectives* and *targets* in ISO 14000.

Note:

a) Write your answer in the space given below.
b) Check your answer with the one given at the end of this book.

____

____

____

____

____

____

____

____

7.4 BENEFITS OF IMPLEMENTING ISO 14001: AN INDIAN SCENARIO

Confederation of Indian Industries (CII) is the nodal agency for ISO 14000 in India. By 2001, over 350 organisations in India have established EMS conforming to ISO 14001. Figure 7.4 gives a sector-wise distribution of these organisations.

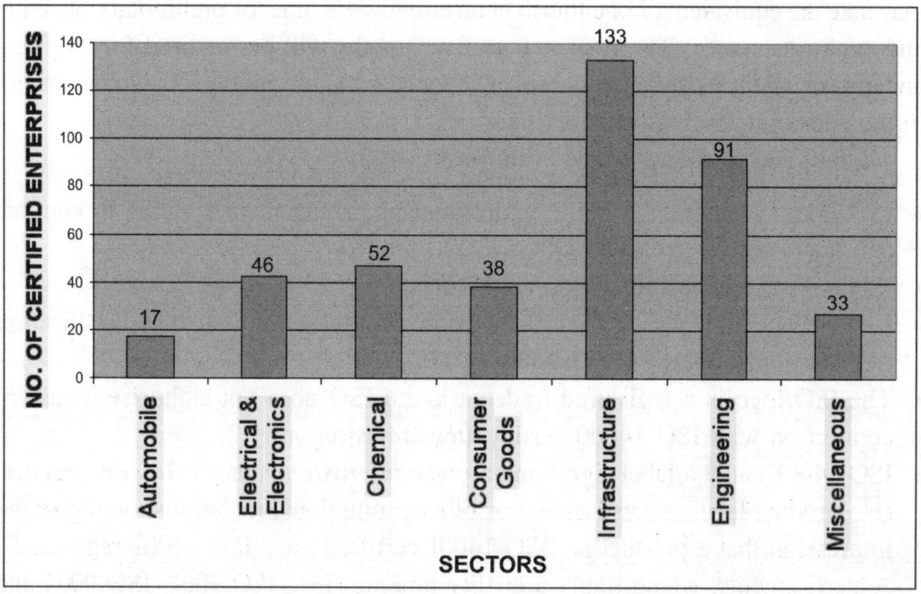

Figure 7.4 Sector-Wise Distribution of 14001 Certified Enterprises

Based on the reports of CII, we will list below the benefits perceived by the organisations after implementing ISO 14001 environmental management system:

- **Environmental benefits:** These include identification of new opportunities for environmental improvements and improvement in resource conservation and waste minimisation.
- **Cost containment and cost savings:** These include identification of new opportunities for cost savings and areas of improvement.
- **Regulatory compliance:** This includes risk reduction and a lower rate of regulatory enforcement.
- **Competitiveness:** This includes an increase in market share and a decrease in cost of production.
- **International trade:** This includes an increase in export-related business opportunities.
- **Greening the suppliers:** These include the opportunities for greening the supplier chain. (Note that this has been a major thrust area for those firms who have a wide range of suppliers. At present, most firms generally do not include

environmental issues in their purchase contracts. Firms have to impart education/ training and transfer of know-how, before actually asking the supply chain *to go green*. Before long, the suppliers' environmental performance will become important for maintenance of quality and compliance of environmental standards.)

- **Other benefits:** These include achievement of commitment to good environmental performance at all levels, safe and healthy working conditions as well as better housekeeping.

Some specific cases of firms that have benefited from the establishment of EMS conforming to ISO 14001 are given below:

- The case of an international electrical and electronic goods giant. It has a very comprehensive and ambitious EMS programme for more than 200 of its factories worldwide supported by the parent body. Other than ISO 14001, it also conforms to a headquarter-driven quality programme. The environmental aspects recognised by the organisation were development and production of different types of capacitors; manufacture of lamps along with their components and manufacture of audio products, PCBs and electronic and electromagnetic ballast. The most important motivation factors were directions from the headquarters; enhancing the organisation's image; resource conservation benefits and ensuring better compliance with legal requirements. After the implementation of EMS, the organisation identified some intangible benefits, which had not been included in its original goal. These benefits were better interaction among suppliers, employees and authorities; improved waste management practices; improved employee awareness and enhanced staff morale; better working conditions and greater domestic share. The major tasks the organisation accomplished after implementation of EMS included 50% reduction in water consumption; 12-25% energy saving depending upon the plant; reduction in packaging mass and reductions in emissions by 95%.
- The case of an organisation was certified to *Responsible Care* before it was certified to the ISO 14001 protocol. The organisation achieved cost savings from pollution prevention, energy conservation and waste minimisation. The savings included recovery of various products (Rs. 5,55,000), reduction in water, fuel, electricity consumption (Rs. 17,10,000), reuse (Rs. 7,00,000), and others. Note that though the organisation followed the Responsible Care norms, it was better able to recognise potential areas for cost savings through the adoption of ISO 14001.
- The case of a mining organisation that maintains the heavy hydraulic machinery/ vehicles used for its excavation work in a workshop situated on top of a hill. As part of the maintenance, oil changes were made and oil was carelessly being

thrown on the land, thus contaminating it. Through the EMS, an organisation-wide oil balance study was initiated and it was found that there was a large gap between the oil received and the waste oil generated by the workshop. On a closer study, it was found that the gap accounted for the oil spill on the land. Consequently, the operating practice was changed and trays were installed to collect the waste oil. From the trays, the waste oil was transferred to drums and were sold. This sale resulted in a revenue flow of Rs. 36 lakhs in the first year of the organisation's EMS implementation.

- The case of a paint-manufacturing firm. At this time, there was no monitoring of the amount of water consumed, and different estimates of water consumption ranging between 250 to 400 m^3/day were available. Through the EMS, the organisation actually found that it was 490 m^3/day. It decided to make a significant resource saving by reducing water wastage. After plugging the water leakage points and controlling the overflow in the cooling tower, water consumption was reduced to 210 m^3/day, i.e., a saving of 280 m^3/day.
- The case of an electrical cable manufacturing organisation. This organisation makes aluminum core cable with an XLPE (cross-linked polyethylene) coating, which is a black hard plastic material. In the process of making this aluminum core, there was the mechanical drawing of the wire from a metal block. This operation generates fine aluminum dust and was inhaled by the workers in the plant (as this metal dust was not contained), which caused respiratory problems and further led to the absenteeism of the workers. Through the EMS, the organisation decided to contain this waste generation by encasing the metal drawing operation. The aluminum dust generated was properly collected and sold to a local paint manufacturer who used it to manufacture silver paint.
- The case of a leading cloth mill. This organisation is the largest denim-manufacturing firm in India. It was the first denim manufacturer in India to gain the ecologically optmised fabric (EOF) trademark, an eco-tex certification. It exports nearly 75%–80% of its total denim production. The most important motivating factors for the firm were enhancing the organisation's image; augmenting management systems culture from quality to other areas of social concern and ensuring legal compliance and achieving higher environmental performance standards. During the implementation phase, the organisation's main focus was on resource conservation. The benefits that the firm realised after the implementation of ISO 14001 include graduation from ad hoc arrangements to using systems approach for environmental management that is consistent with the overall operations of the unit, incorporation of environmental parameters in the corporate decision-making process and enhancement of the organisation's image and greater worker motivation.

LEARNING ACTIVITY 7.5

State how an industry benefits from an EMS.

Note:

a) Write your answer in the space given below.

b) Check your answer with the one given at the end of this book.

SUMMARY

In this Unit, we explained how EMS standards help organisations establish a structured process of continuous environmental improvements that are monitored, measured and reported. We began the Unit with a discussion of the core elements and benefits of EMS. We also touched upon the registration process for obtaining the environmental certification. We then traced the genesis and evolution of ISO 14000 Series and discussed the principles and structure that govern the Series. Following this discussion, we focussed on EMS specification standards, i.e., ISO 14001 Series, and explained the process involved in the implementation of ISO 14001 series.

REFERENCES

Global Green Standards: ISO 14000 and Sustainable Development, International Institute for Su s t a i n a b l e Development, http://www.iisd1.iisd.ca

Tech Monitor, Special Issue on Environment Management, Vol. 12, No. 5 (Sept–Oct 1995)

SUGGESTED READINGS

Cascio, J. (ed.), 1996. *The ISO 14000 Handbook*, CEEM Information Services, 764p.

Global Green Standards: ISO 14000 and Sustainable Development. International Institute for Sustainable Development.

Kanholm, J (ed.), 2000. EMS Manual, 21 Procedures, and Forms. AQA Press, USA.

National Safety Council. *Industrial Safety Chronicle,* Special issue on ISO 14000, Sept 2001.

Sayre, D., 1996, *Inside ISO 14000: The Competitive Advantage of Environmental Management*, St Lucie Press, Delray Beach, FL

Strategic Environmental Management: Using TQEM and ISO 14000 for Competitive Advantage (Wiley Series in Environmental Quality Management), by Grace H. Wever, April 1996.

The ISO 14000 Essentials: A Practical Guide to Implementing the ISO 14000 Standards, canadian Standards Association.

Willig, J.T., 1995. Auditing for Environmental Quality Leadership*:* Beyond Compliance to Environmental Excellence. *New York, NY: John Wiley and Sons, Inc.*

ANSI (American National Standards Institute). *http://www.ansi.org*

BSI (British Standards Institution). *http://www.bsi.org.uk*

EPA Standards Network. *http://www.epa.gov/partners/iso*

ISO (International Organization for Standardization). *http://www.iso.ch*

ISO 14000.com Home Page. *http://www.iso14000.com*

ISO 14000.org Home Page. *http://www.iso14000.org*

UNIT 8

Environmental Management Techniques

STRUCTURE

OVERVIEW

In Units 4 to 7, we discussed environmental management tools. In Unit 8, we will discuss some related environmental management techniques. We will begin the Unit by discussing environmental monitoring. We will then explain how modelling helps in applying the environmental tools in real-life situations for quantification of impacts, prediction of scenarios and simulation studies. We will also discuss other techniques such as sensitivity analysis, remote sensing, environmental profiling, environmental technology and risk assessments and eco-mapping. We will then introduce you to the techniques that can be adopted in the urban contexts. Finally, we will discuss the importance of environmental education, i.e., a learning process that increases people's knowledge and awareness about the environment and associated challenges, develops the necessary skills and expertise to address the challenges and fosters attitudes, motivations and commitments to make informed decisions and take responsible actions.

LEARNING OBJECTIVES

After completing this Unit, you should be able to:

- state the processes involved in monitoring, modelling, sensitivity analysis, remote sensing, environmental profiling, technology assessment, risk analysis and social impacts;
- use these techniques where relevant;
- carry out eco-mapping;
- discuss the need for and the role of environmental education in environmental management.

8.1 ENVIRONMENTAL MONITORING

Environmental studies often require information on physical, chemical, biological, economic or social aspects of particular environments. These can be obtained from a monitoring programme involving surveys, measurements and data collection activities. A monitoring programme, thus, helps establish baseline information and data for describing the present situation of an area likely to be impacted by a proposal.

Monitoring may involve sampling of air, water, vegetation and soil, and these data collection programmes should be planned to obtain the greatest value from the data, which is often expensive to collect and process. Care should be taken to classify, analyse and store data for easy retrieval, so that it can be useful as baseline or reference data for other assessments.

In environmental management, monitoring has a role to play even after the completion of environmental studies and the subsequent implementation of the projects. In many cases, for example, some impacts may remain uncertain, as there may be insufficient time or resources to assess these with complete confidence. If when applying the precautionary principle, the likely adverse effects of a project appear to be low and to be outweighed by the benefits, a regulatory authority may permit the project to proceed, but with a requirement for monitoring the effects.

The owners of the project are, therefore, obliged to carry out a monitoring or surveillance programme continuously or at defined intervals, and to report the results to the regulatory body. If adverse effects beyond those anticipated in the original environmental impact assessment become apparent, remedial actions must be taken. However, in situations where suitable remedies are not available, it is possible that the projects are terminated.

A post-implementation monitoring of a project may involve audits such as:

- Implementation audits to determine whether the recommendations or requirements in an EIS were implemented.

- Project impact audits, to determine the actual impacts of a project, independent of the predictions made.
- Predictive techniques audits, to assess the predictions made in the EIS and the methods of prediction used by comparing actual outcomes with forecasted ones (This will aid future studies).

Before you read further, note the distinction between monitoring, survey and surveillance programmes given below:

- **Monitoring:** A long-term, standardised measurement programme involving observation, evaluation and reporting of a part of the environment in order to define status and trends.
- **Survey:** A finite duration, intensive programme to measure, evaluate and report the quality of part of the environment for a specific purpose.
- **Surveillance:** A continuous, specific measurement programme involving observation and reporting for the purpose of environment management and operational activities.

8.2 ENVIRONMENTAL MODELLING

Before we discuss environmental modelling, let us first explain what modelling means. A model is a representation of real-life problems or situations. It copies significant attributes of a real prototype but is simpler and is easier to build, change or operate. Put differently, models are basic tools in science, engineering, business and various forms of planning. They can be applied to situations or systems, which are both existing and non-existent. For example, suppose that we are to assess the environmental impacts of releases of heated water from a proposed thermal power plant into a shallow lake. To assess the impact, do we have to build a power station and deliberately release varying amounts of heated water, as an experiment? Given the huge efforts, great costs involved, etc., this is unwise to do. Furthermore, this will cause the very environmental damage that the planning exercise is trying to prevent. To assess impacts, it is obviously better to use a model of the system.

Essentially, models allow us to extrapolate from the existing systems and knowledge to analyse potential situations. They are only useful to the extent that they accurately model the real world. Models can be constructed from logic or rational assumptions, scientific theories and information about similar situations or operations.

The model structure may be set up to search for an optimal answer (as in linear or dynamic programming models) or to generate possible solutions (e.g., stochastic models). The great majority of models, however, just stimulate the behaviour of a system. Variations can be explored by running the model several times, changing inputs or other features. There are many kinds of models such as statues, model aeroplanes, scientific theories, hydraulic laboratory models and computer programmes. We can classify models into the following:

- Iconic models, which have similar attributes to the prototype, e.g., a model-racing car.
- Analogue models, where some aspects of the model are analogous to the prototype, although they differ physically, e.g., flows of heat, electricity and groundwater act similarly, and electric analogues have been used to model groundwater flows.
- Symbolic or mathematical models, which describe physical situations using abstract representations such as diagrams, equations, sets of calculations or computer programmes.

Mathematical Model

The most widely used types are mathematical models, which use logical relationships, rules, variables, equations, parameters and algorithms (i.e., sets of linked equations). These describe systems, made up of various parts and connections, rather than self-contained entities. Figure 8.1 illustrates a simple mathematical model.

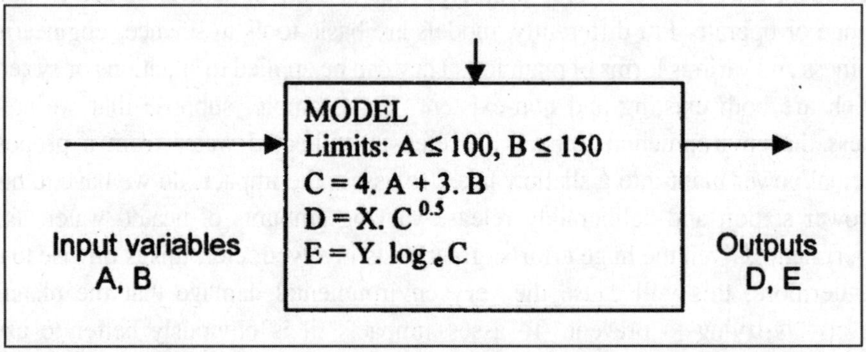

Figure 8.1 Simple Representation of Mathematical Model

The mathematical model of the type depicted in Fig. 8.1 above may describe processes such as:

- Pollution of an area of land by chemicals carried by groundwater, with pollutant and water quantities as inputs, soil properties as parameters, pollutant concentrations at various locations as outputs and a mathematical description of groundwater flows as the model. (Calculations will proceed for various time steps, say, at daily intervals, and, therefore, there will be sets of input variables, rather than single values.)

- A cash flow analysis with receipts and payments as inputs, discount rates as a parameter, net present value as an output and compound interest equations as the model.
- The effects of development and environmental degradation on land values, with development and pollution levels as inputs, indices of social reactions as parameters and land values as outputs and relationships involving benefits of development, disadvantages of pollution, public perceptions and land values as the main model.

Other examples of the above are oxygen sag in a river, operation of an industrial plant, econometric models describing consumer responses to changes in energy prices and ecological changes in an altered environment.

You must note that:

- Models can be much more complex than the simple structure shown in Fig. 8.1, with iterations and feedbacks. A steady-state model can describe static situations or snapshots of a system at a particular time. Situations changing with time can be modelled by unsteady or dynamic models, which work with a series of time steps. These usually employ the calculated outputs at the end of one time step as inputs at the beginning of the next step. Probably, the most famous models are those used by the Club of Rome (Meadows et al., 1972, Meadows et al., 1992), which stimulated the future of the earth, exploring global scenarios of population growth, industrial production and environmental degradation.
- Models may be purpose-built for a particular application, or general models which often use proprietary computer programmes. Examples of general models are those describing storm water pollution transports, such as the Storm Water Management Model (SWMM) from the U.S. Environmental Protection Agency and the Storage, Treatment, Overflow Runoff Model (STORM) from the U.S. Army Corps of Engineers. When general programmes or packages are used, you should realise that two models are involved—the standard computer model which provides a *shell* into which a model for a particular situation (expressed in the input data file) can be fitted.
- While most computer models were written in languages such as FORTRAN in the past, many modellers are now likely to use spreadsheet programmes such as Lotus 1-2-3 and Excel. These offer facilities for basic stimulation, special functions such as random number generation for Monte Carlo Analysis, regression and other statistical tests and easy presentation of results as charts

and graphs. Microcomputers provide adequate computer power for most modelling applications.

- Models are ineffective without data and calibration. Model results are sometimes accepted without adequate scrutiny because they are generated through a computer. The axiom *rubbish in – rubbish out* applies to all computer programmes. Computer generated results must therefore analyse critically and should be checked for consistency and logic, and if possible, validated against additional data.

The accuracy of a model depends upon the following, in order of importance:

- amount of data used to build and operate the model;
- experience or skill of the analyst;
- quality of the model.

Modelling exercises involve prediction of uncertain system behaviour, usually during some future period. Forecasting is, therefore, necessary in environmental management, as it helps in prediction of future periods. Growth modelling is required in areas such as estimation of future population for urban area, growth of fauna in forest, etc. We will discuss these two types in Subsection 8.2.1 and 8.2.2, respectively.

LEARNING ACTIVITY 8.1

List the input variables, system parameters and output parameters you would consider for developing a model for the estimation of the impact of industrial sources of noise on nearby habitations.

Note:

a) Write your answer in the space given below.

b) Check your answer with the one given at the end of this book.

8.2.1 Forecasting Modelling

It is important for you to note that the scope of forecasting techniques is limited. For example, the longest time for a reasonable population projection is one decade, whereas for economic projection the timeframe is, at the most, 12 months. Long-term forecasts by expert think tanks are usually wide off the mark. In addition, all forecasts can be invalidated by some drastic event, such as a war or a natural disaster. Thus, we should have a healthy skepticism about forecast results, and apply checks on them.

The factors/data needed, among others, to make a forecast are:

- Information on the situation being forecast, preferably past records of the main factor of interest and related factors.
- A model of the system or situation, which could be a simple concept, such as the input-output system in Fig. 8.1, or a full scientific model based on physical description and theory, such as the computer models used in weather and climate prediction.

Forecasting can be done on the basis of experience of judgement, but it is better to have some explicit model or basis. These can be causal or explanatory, seeking to describe the processes producing the phenomenon being predicted, or non-causal, considering results numerically or statistically, but not trying to define underlying processes.

Figure 8.2 shows the forecasting problem for a particular variable using past record of values for extrapolation.

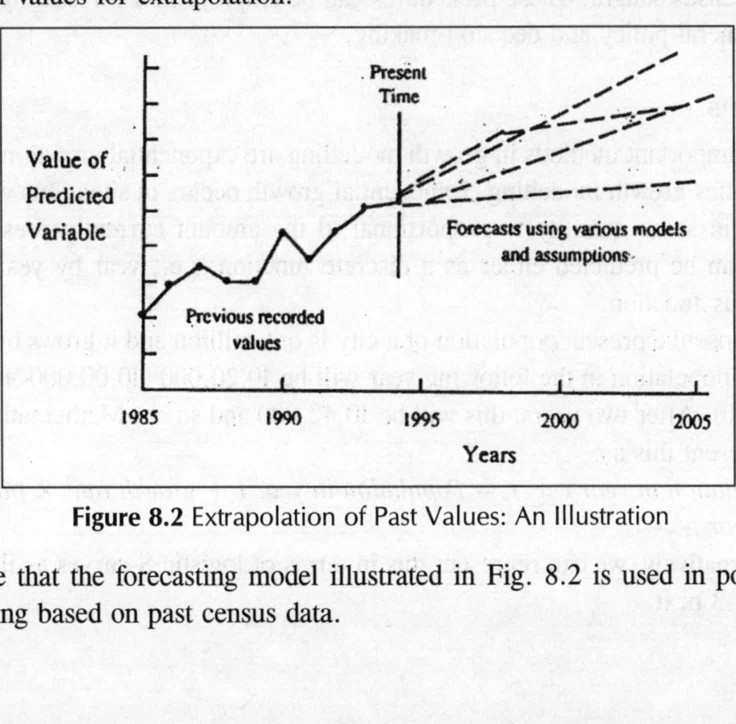

Figure 8.2 Extrapolation of Past Values: An Illustration

Note that the forecasting model illustrated in Fig. 8.2 is used in population forecasting based on past census data.

A systematic forecasting may use such mathematical models as linear, multiple and non-linear regression relating future values to the past record or time series of values, smoothing using simple and double moving averages, exponential smoothing, etc., filtering and Box-Jenkins.

The literature is replete with mathematical forecasting models. However, to use these models properly, it is necessary to set up databases of past records.

Note that simple regression techniques are useful to establish a relationship between the value and time, which can be extrapolated to make predictions. Smoothing techniques help reduce the variability of data. Moving averages, for example, replace each value with the mean of this value and the values occurring immediately before or after it, and this allows us to work with a reduced variable set of values for extrapolation. Thus, we may average 3, 5 or more values. Spreadsheet programmes for computers now offer facilities for regression, time series analysis and smoothing.

When data availability is limited, one of the right models is the cause-effect model. Causal model include factors, which are known to influence the variable being considered. For example, the quantity of rice sold by an agricultural co-operative at the end of a harvest may be related to the cost of the fertiliser, the price of rice, the price of wheat (a competing product) and the rainfall for the growing season.

An alternative to mathematical modelling is to seek the opinions of experts. The Delphi technique and other procedures combine the views of experts in a series of structured meetings where alternatives and scenarios are presented, debated and consensus sought. These procedures can be employed for forecasting, as well as for general policy and decision-making.

8.2.2 Growth Modelling

The two important methods in growth modelling are exponential growth modelling and logistics growth modelling. Exponential growth occurs in situations where the increase in some quantity is proportional to the amount currently present. The growth can be predicted either as a discrete function, i.e., year by year or as a continuous function.

Suppose the present population of a city is one million and it grows by 2% per year, the population in the following year will be 10,20,000 (10,00,000 + 0.02 × 10,00,000). After two years, this will be 10,42,000 and so on. Mathematically, we can represent this as:

Population in year t + 1 = Population in year t + growth rate × population in year t.

Alternatively, we can represent this in terms of logistic S-curves as illustrated in Fig. 8.3 next.

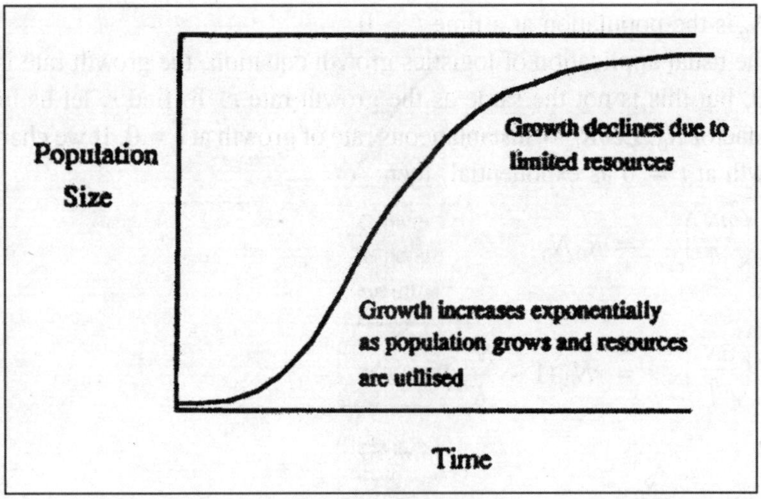

Figure 8.3 Logistic S-Curve as a Limitation on Populations

Biologists successfully use logistic curve models to model populations of many organisms including protozoa, yeast cells, etc. Such models describe the rate of change in population as a function of present population, natural growth rate and carrying capacity of the ecosystem. Mathematically, this is represented as:

$$\frac{dN}{dt} = rN(1 - \frac{N}{K})$$

where N is population size; r is growth rate and K is the carrying capacity of the environment.

Notice that when N is less than K, the rate of change of population is proportional to the population size. That is to say, the population grows exponentially with growth rate r. As N increases, the rate of growth slows down, and eventually, as N approaches K, growth stops altogether and the population stabilises at a level equal to the carrying capacity. The factor $(1-\frac{N}{K})$ is often called the environmental resistance, i.e., as the population grows, the resistance to further population growth continuously increases.

The solution to the above equation is:

$$N = \frac{K}{1+e^{-r(t-t^*)}}$$

Note that t^* corresponds to the time at which $N = K/2$. Substituting $t = 0$, lets us solve for t^*:

$$t^* = \frac{1}{r} ln\left(\frac{K}{N_0} - 1\right)$$

where N_0 is the population at a time $t = 0$.

In the usual application of logistics growth equation, the growth rate is known at $t = 0$, but this is not the same as the growth rate r. To find r, let us introduce another factor R_0. Let R_0 = instantaneous rate of growth at $t = 0$. If we characterise the growth at $t = 0$ as exponential, then:

$$\left(\frac{dN}{dt}\right)_{t=0} = R_0 N_0$$

But,

$$\left(\frac{dN}{dt}\right)_{t=0} = rN_0\,(1 - \frac{N_0}{K})$$

And so,

$$r = \frac{R_0}{1-\left(\frac{N_0}{K}\right)}$$

Using this equation, we can use quantities that are known at $t = 0$, namely the population size (N_0) and the population growth rate (R_0), to find appropriate growth factor (r).

The logistic curve can also be used to introduce another useful concept in population biology called the *maximum sustainable yield* of an ecosystem. The maximum sustainable yield is the maximum rate that individuals can be harvested (removed) without reducing the population size. Let us consider fish harvesting as an example. If the fish pond is at its carrying capacity, there will be no population growth, and, therefore, any fish removed will reduce the population. In other words, the maximum sustainable yield corresponds to some population size less than the carrying capacity. In fact, since the yield is the same as dN/dt, the maximum yield corresponds to the point on the logistic curve where the slope is at a maximum. We can find that point if we set the derivative of the slope equal to zero. The slope of the logistic curve is given by:

$$\frac{dN}{dt} = rN\,(1 - \frac{N}{K})$$

Setting the derivative equal to zero gives:

$$\frac{d}{dt}\left(\frac{dN}{dt}\right) = r\frac{dN}{dt} - \frac{r}{K}\left(2N\frac{dN}{dt}\right) = 0$$

This yields:

$$1 - \frac{2N}{K} = 0;\ N = \frac{K}{2}$$

This means if the population growth is logistic, then the maximum sustainable yield will be obtained when the population is half the carrying capacity.

In any analysis modelling, certain degrees of uncertainty are unavoidable. However, by carrying out sensitivity analyses, we can determine the degree of uncertainty. Let us discuss it next. But, first, work out Learning Activity 8.2.

LEARNING ACTIVITY 8.2

Suppose the human population follows a logistic curve until it stabilises at 15.0 billion. In 1986, the world's population was 5.0 billion and its growth rate was 1.7%. Calculate when the population will reach 7.5 billion and greater than 14 billion.

Note:

a) Write your answer in the space given below.

b) Check your answer with the one given at the end of this book.

8.3 SENSITIVITY ANALYSIS

Any analysis, be it an environmental model, a cost-benefit study or other investigation, will involve a number of input factors having different degrees of uncertainty. These will influence the outcomes of the study to varying extents. In an input-output process (see Fig. 8. 1), we usually select the inputs as the most-likely values. The relative responses of outputs to changes in inputs are termed by their sensitivity. That is to say, if large changes in an input produce insignificant changes in an output, the output is insensitive to the input.

A basic test of sensitivity is whether a percentage change in an input factor produces a higher or lower percentage change in an output. For example, consider now the model shown in Fig. 8.1. The factors in the analysis can be seen as inputs (*A, B*), and the outcomes as outputs (*D, E*). If the parameters were set at $X = 1.1$ and $Y = 1.9$, and the most likely inputs were $A = 75$ and $B = 102$, the outputs will be $D = 27.1$ and $E = 12.2$.

Now, suppose that input *A* is considered to be accurate to ±40%, and input *B* to be accurate to ±25%. The limits for *A* may, therefore, be from 45 to 105, and for *B* from 76.5 to 127.5. Taking the highest and lowest sets of values, we can repeat the calculations. The low values of *A* and *B* lead to outputs of $D = 22.3$ (–18%) and $E = 11.4$ (–7%). The high values of inputs (with A truncated from 105 to 100) will give $D = 30.8$ (+44%) and $E = 12.7$ (+4%). The percentage figures show that the outputs are relatively insensitive to the 25% changes in inputs. This is due to the square roots and log functions in the equations for *D* and *E*. Other relationships may amplify the changes in inputs and accordingly give sensitive responses.

The extent of the change in an outcome, such as a level of CO_2 emissions or a net present value, NPV, relative to the change in the input, can be used as a mathematical indicator of sensitivity, similar to the concept of elasticity in microeconomics. Some mathematical programming procedures (e.g., linear programming) calculate sensitivity factors.

Sensitivity analysis is a powerful, yet simple, technique for determining the effects of individual factors and their variations on the overall results of an analysis. It can be applied to any analysis, which can be visualised as system inputs and outputs, and merely involves the repetition of calculations. This process is greatly facilitated, if all or part of the analysis can be set up on a computer. Spreadsheet programmes (Excel and Lotus 1-2-3) derive their popularity from their ability to carry out sensitivity studies with ease.

In financial studies, for example, the numbers can be set out in a grid to represent a cash flow or a spending budget, and indicators such as NPVs or gross profits can be calculated by the formulae associated with particular grid locations. The alteration of any number in a grid causes all values determined by formulae to be re-calculated. Similar calculations can be performed with models of other processes. In short, the most basic use of sensitivity analysis is exploring "What if?" situations.

Note that there is no formal procedure for sensitivity analysis. For example, it can be applied by:

- examining factors one at a time and determining the variation in outcomes due to changes to a single factor, keeping the others at their most likely values;

- taking the best or worst estimates of all factors to see how a system performs under extreme conditions (If it performs favourably when all factors are set at their most unfavourable values, the analyst can be very confident about outcomes.);
- varying a factor sufficiently to cause a reversal of the outcome given with its most likely value. (For example, if a transportation project produces pollution in excess of some regulatory standard, the analyst could determine how far the process must be modified to meet the standard. This technique requires an iterative search, and is probably only practical if an analysis has been computerised.)

It is useful to carry out sensitivity analyses at a preliminary stage of a large study to identify which of the factors involved have the greatest bearing on results. Particular attention can then be paid to data collection and estimation, so that they can be estimated as accurately as possible. The less important factors need only be estimated approximately.

LEARNING ACTIVITY 8.3

In the prediction of impacts of combustion of coal in thermal power plant on surrounding air quality, describe three parameters which you would consider in sensitivity analysis.

Note:

a) Write your answer in the space given below.

b) Check your answer with the one given at the end of this book.

Following our discussion of sensitivity analysis, which is important in any kind of terrain study, we will next discuss the application of remote sensing and GIS in environmental management. Modelling and sensitivity analysis find immense use in such applications.

8.4 APPLICATION OF REMOTE SENSING AND GIS IN EM

Benefits of harnessing the new developments in high technology areas like space technology and information technology for sustainable development have been well recognised and many developing countries are looking towards assimilating these technologies as part of their developmental plans. Satellite remote sensing integrated with Geographical Information System (GIS) technology provides a tool for addressing the issues of spatial reference in enhancing the quality of life and sustainable development.

Geographical Information Systems are computer aided decision support and planning tools, which integrate data from maps (spatial data) and other auxiliary data (attribute data) for a geographical area of interest. They can be used to create and maintain geographic databases and are eminently suited for what-if analysis in any planning related activity. GIS applications are developing rapidly. A GIS application is able to provide many simulated results, which help in making informed decisions. These simulated results also help decision-makers in addressing such questions as the following:

- Where is the most polluted area in the industrial area and how much is the total?
- Where is the exact location of the polluted rivers and total number of industrial units along the river?
- How many premises are violating the environmental norms?
- Where are the projects located that need to be considered in an environmental impact assessment (EIA)?

The basic advantage of the GIS technology is its ability to manage and integrate with the existing database. This means that at any point in time we are able to see not only the spatial data but also the database. For example, by referring to the colour codes displayed within a map on a GIS application, we will know that grey area is a polluted area and the green area is free from pollution. At the same time, the presentation of database related to the spatial data can also be shown.

The implementation of GIS for the purpose of controlling and monitoring development involves the stages of data gathering and updating; development of GIS database; development of user interface and application of GIS database.

Once the GIS package integrates the spatial and attribute data, the planner has a powerful tool, which can be used for information dissemination or analysis. By querying the geographic database in several ways, the planner can present the available information in a variety of formats (e.g., printed tabular reports, graphically as map display and map outputs on paper). The analysis capabilities of a GIS package allow the planner to address what-if questions and work out a variety of action plans in a scientific manner.

A number of problems can be solved by geographic analysis. Typical examples include township development; relationships between agricultural parameters (e.g., yield and salinity), land capability analysis, site locations for facilities and environmental problems (animal migration).

Remote sensing and GIS technologies play a vital role in urban and regional planning, resources monitoring and allocation and environmental management. India is one of the few countries in the world that has made remarkable progress in space technology. India has launched a series of operational remote sensing satellites providing high quality remote sensing data to several parts of the world. By integrating remote sensing and GIS methodologies, India has carried out a number of operational projects at national level in the areas of resources planning and management. In response to rapid developments in computer hardware, graphic display systems and spatial theory, GIS technology is achieving wide spread popularity among the administrators, managers, line department personnel and academia.

In Sections 8.5 to 8.9 we will discuss some environmental tools that can be used in urban contexts.

8.5 ENVIRONMENTAL PROFILE

Life in cities depends upon a wide variety of development activities such as industry, commerce, transportation, construction, households, etc. These activities depend upon, as well as damage, environmental resources such as land, water, air and ecological systems.

Environmental profile (EP) is normally the first project activity to be undertaken for a city level project implementation. It brings together information about the city's development sectors and activities, its environmental resources and hazards and management systems in a way that allows a systematic analysis of how development and environment interact. In this way, EP not only highlights and elaborates the key environmental issues facing the city, but also puts them equally into the appropriate development and management context. It also identifies the different groups, organisations and stakeholders, which have important (and often conflicting) interests.

Two main purposes of EP are:

(i) To provide a systematic overview of a city's development activities and how they interact with the city's environmental resources.

(ii) To identify and mobilise stakeholders by being a source of relevant information and a process of its preparation.

The four main chapters of EP are city introduction, development setting, environment setting and environmental management. We will discuss these, next.

City Introduction

This should give a highly summarised and selective introductory information about the city including its geography and physical setting, social characteristics and economy. The purpose of this chapter is to provide information about the city, which is relevant and important for understanding the other chapters that follow. This chapter includes one or two A4 sized basic maps.

Development Setting

This chapter examines environment development relationships from the point of view of development activities. It discusses the development activities of the city (e.g., manufacturing, mining, fisheries, transport, housing, etc.) The main purpose is to describe the use of environmental resources by each activity sector in terms of type, quantity and quality of resources used. For example, local manufacturing industry may consume groundwater as part of its production process and may also use waterways and the atmosphere for getting rid of the waste. This chapter also analyses the impacts of each sector on environmental resources and examines environmental hazards.

Environment Setting

This chapter examines the city's environmental resources and the ways in which the city's activity sectors make use of and have impacts on these resources. In the context of the EP, environmental resources are primarily those things arising from the physical world, which are used to support urban life and development. This means air, water, land (and the living and non-living organisms in them) as well as areas of special ecology and special human-made environments such as cultural and historical buildings. This chapter analyses each of the environmental resources separately in the following five sub-sections:

(i) **Characteristics of the resource:** This Subsection describes the resource in terms of quality, quantity and different ways in which it is used.

(ii) **Use of resource by activity sectors:** This Subsection discusses the use of the resource by all activity sectors together with respect to level of extraction, consumption and depletion.

(iii) **Impacts on the resource from activity sectors:** This Subsection discusses the impact of the activity sectors on the resources, including impacts of pollution and damage.

(iv) **Competition for use of the resource:** This Subsection discusses the conflict among the activity sectors in the use of the resource.

(v) **Management arrangements:** This Subsection discusses institutional arrangements or initiatives, including the organisations responsible, special arrangements such as commissions, projects, etc., to help manage the resource.

Environment Management Setting

This chapter has three main sections: key stakeholders, urban management structures/institutions and strengthening urban and environmental management. The purpose of the first section (i.e., key stakeholders) is to identify and discuss only those stakeholders that play significant roles in urban development and urban environmental management. The stakeholders can be classified into the public sector, community sector, private sector and other interest groups not covered by the other groupings. When describing the various stakeholders, the chapter generally gives a brief account of their activities and roles. This account pertains to the information, knowledge and technical expertise they possess or have access to; the extent to which they are involved in decision-making, policy formulation and policy coordination and their involvement in the implementation of the city's development and environmental policies, programmes and projects.

The second section (i.e., urban management structures/institutions) gives an overview of the city's organisational structure using four headings, i.e., basic organisation and structure; information, knowledge and technical expertise, and how it is applied and shared; process of decision-making, policy formulation and coordination and policy implementation. The third section (i.e., strengthening urban and environment management) discusses current or planned initiatives and efforts to strengthen the city's urban and environmental management systems and to increase local abilities to plan, coordinate and manage sustainable urban development.

In Section 8.5, we described EP as a tool that brings together information about a city's development sectors and activities, its environmental resources and hazards and management systems in a way that allows a systematic analysis of how development and environment interact. In Section 8.6, we will describe another tool, i.e., environmental technology assessment for urban contexts.

8.6 ENVIRONMENTAL TECHNOLOGY ASSESSMENT

An environmental technology assessment (EnTA) is an analytical tool used to help understand the likely impact of the use of new technologies in terms of costs, monetary benefits, environmental effects and social and political impacts. It focusses on the analysis of effects on the environment, specifically on human health, ecological systems and resources. It is explored as an analytical tool, designed to ensure that decision-making processes related to technology adoption, implementation and use are sustainable.

Decision-making is normally tiered in a sequence of multiple decisions. Multiple decisions often progress from a strategic or policy level, through plans and programmes to specific project level decisions. It is argued that more conventional analytical tools such as environmental impact assessment (EIA) and social impact assessment (SIA) fail to address issues at the higher tiers of decision-making and that an EnTA, a flexible approach to technology related decision-making, is well positioned to fill this gap.

Potential Users and Scope of EnTA

The informal voluntary nature of EnTA is most appropriate at the politico-strategic decision-making levels within the urban management context. As a strategic decision-making philosophy that enhances sustainability rather than a prescribed process, an EnTA is aimed at administrators providing strategic and policy level guidance.

Hierarchical Decision-making at Local Authority Level

Decisions on technology adoption, implementation and use at local authority levels are often one-shot events. Instead, they are characterised by a multiple sequence of decisions made across a tiered hierarchy, a process that is called *disjointed incrementalism*. The decision-making sequence starts with conceptual or strategic policy type decisions at the apex, followed by semi-conceptual or semi-strategic plans and programmes, culminating in detailed or specific project related decisions as shown in Fig. 8.4.

As illustrated in Fig. 8.4, decision-making at the apex is often confined to broad, strategic, policy type decisions, which are frequently politicised. While inadequate information may increase the uncertainty level, which is usually characterised by time constraints, information requirements are limited to conceptual and generic directions. More often than not, decision outcomes represent policy statements that are intended to give direction to lower tier decisions. Critical

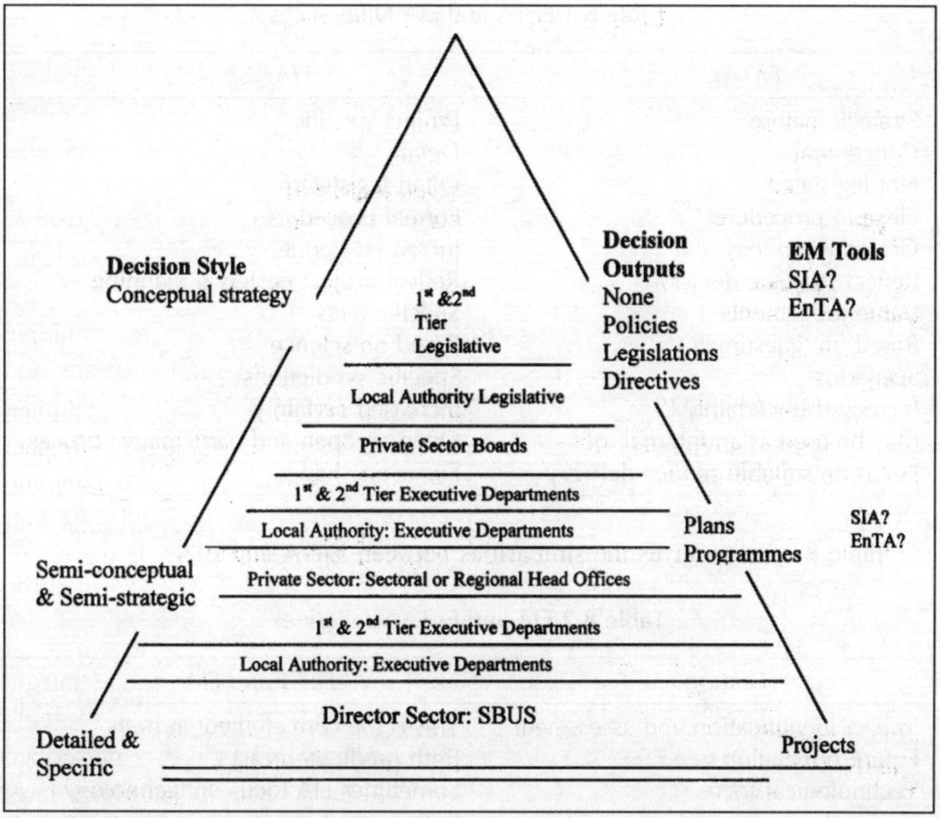

Figure 8.4 Specific Project Related Decisions

concerns regarding the relationship between assessment tools and decision-making at the top tiers include the following:

- EIAs are seldom done for decisions at this tier of decision-making.
- EIAs are not designed to meet the challenges of decision-making at this tier in part due to time constraints.
- EIAs are often commissioned too late to address issues that have been resolved at the top tier.
- EIAs are very seldom used to challenge policy level decisions with the result that many issues are foreclosed when they are conducted.

Comparison of EnTA and EIA

A comparison between EnTA and EIA explores both differences and similarities. Table 8.1 gives the differences between EnTA and EIA.

Table 8.1 EnTA and EIA Differences

EnTA	*EIA*
Strategic nature	Project specific
Conceptual	Detail
Not legislated	Often legislated
Flexible procedure	Formal procedures
Greater simplicity	Increased rigour
Reflect strategic decision	Reflect project design & planning
Value judgements	Specific facts
Based on 'guestimates'	Based on science
Scenarios	Specific predictions
Increased uncertainty	Increased certainty
May be used as an internal tool	Often an open and participative process
Focus on suitable service delivery	Focus on change

Table 8.2 below gives the similarities between EnTA and EIA.

Table 8.2 EIA and EnTA Similarities

Feature	*EnTA and EIA*
Impact identification and assessment	This is the core element in both
Future orientation	Both predict/forecast
Technological focus	Sometimes EIA focus on technology
Comprehensive	Both aspire to be as comprehensive as possible
Scale	This is a variable in both

Benefits of EnTA

EnTA is a flexible, informal and voluntary tool that defines an appropriate approach to sustainable decision-making. EnTA should be used to ensure increased eco-efficiency and sustainability of both technologies. Eco-efficient technologies or environmentally sound technologies perform better across all the dimensions influencing sustainability of technology selection, adoption and use, as well as the various phases of the project execution cycle. Improved eco-efficiency of technologies may be evaluated in terms of parameters such as improved environmental sustainability (i.e., improved resource and energy consumption, reduced emissions and effluents, reduced wastes and improved safety and health profiles), increased net social benefit, equity and empowerment, improved economic sustainability, improved financial sustainability, improved technical sustainability and improved institutional capacity.

Major In-roads and Applications

It is not possible to report on specific EnTA case studies, as EnTA is primarily a voluntary and internal tool that is used by organisations to ensure that technology related projects are guided by sustainable policy decisions before more project specific environmental impact assessment (EIAs), hazard and operation ability studies (HAZOPS) or risk assessments are done. However, an entity like the Development Bank of South Africa has done EnTA-like analysis of prospective borrowers for implementing water treatment technologies. A large petrochemical company has also internalised hybrid EnTA and EIA principles into their project appraisal procedures.

Note that because EnTA, as a philosophy or an approach to technology adaptation processes, is internalised into the decision-making process, it is not documented in the same manner as EIAs, which have a distinct approval phase.

8.7 ENVIRONMENTAL RISK ASSESSMENT

Risk may be defined as the combination of the probability of occurrence of an event and the possible extent of its adverse effects and consequences in terms of economic loss or human injury. Risks can relate *to hazards to humans* from natural disasters, and events such as traffic accidents and diseases and chronic dangers such as cigarette smoking. Health risks such as probabilities of contracting a particular disease are environmental hazards in the sense that they arise in our surroundings. They may be connected with projects such as development of industrial plants or particular production processes, and must be included in environmental assessments. Economic risk is yet another type of risk. Fairly sophisticated mathematical techniques have been developed for analysing economic risks and introducing probabilistic risk assessments into cost-benefit analysis. Other forms of risk may relate to hazards from industrial processes, safety of workers and effects of emissions on natural systems, flora and fauna. Ecological risk is handled using similar techniques as human health risks.

Note that risk involves both uncertainty and severity of damage. There will be no risk, if one knew the exact time, place and nature of an adverse event, and there would be no risk, if no damage occurred. Mathematically, risk can be expressed as:

$$Risk = Probability\ of\ some\ event \times seriousness\ of\ its\ consequences.$$

Risk analysis and risk management have been widely used in insurance and business decisions for many years, and more recently, as a basis for environmental decision-making. Risk analysis is a systematic use of available information to determine how often specified events may occur and their likely consequences.

Its purpose is to identify the causes, effects and magnitudes of risk and to provide a basis for risk assessment and treatment. Risk management is the systematic application of managerial policies, procedures and practices to the tasks of identifying, analysing, assessing, treating and monitoring risks.

Having said that, let us now focus on risks related to industrial production processes.

8.7.1 Environmental Risk Management in Industry

Environmental risk management in companies is a fairly recent phenomenon. Generally, risk management has been associated with insurance and the recovery of financial losses in the event of unexpected disasters. But, catastrophic accidents have increased in number in the last decade, dramatically underlining the fact that there is a serious and growing worldwide problem that must be faced.

Depending on the timing of the damage, risks can be immediate or delayed. The risk can also be *voluntary* (e.g., smoking, driving, etc.) or *involuntary* (e.g., living close to a potentially dangerous industrial plant). Furthermore, risks can be represented in two different forms, namely, individual risk and societal risk. Individual risk is the frequency with which an unprotected person is affected by a specific type of damage. Societal risk is the relationship between the frequency of an accident and the number of people injured or killed.

A situation of risk can involve an opportunity for success as well as a prospect of failure. If the risk can result in failure only, it is a *negative risk* and is usually defined as static (e.g., the risk of a fire in a company's properties). A risk that can result in both success and failure is defined as a dynamic risk (e.g., buying shares in the stock market). Environmentally related risks are mostly limited to situations where an event can only cause environmental degradation and/or adverse human health effects, and they must therefore be considered as static risks.

Companies are built and managed on the whole concept of risk. No matter how big or small a business decision may be, it is based on the degree of risk for success or failure. Risks in companies can find expression in many different ways (e.g., risks of fire, industrial injuries, product responsibility, industrial espionage and environmental disasters). However, the two crucial problems that usually arise when dealing with risks are the following:

(i) To what extent can they be considered the maximum acceptable level?
(ii) To what degree should these risks, considered acceptable, be further reduced and how should a reduction be balanced against costs and societal constraints?

For companies, many environmental risks can also be associated with unexpected events, like industrial accidents, or with routine events such as permitted emissions to air and waterbodies. Some of these more generalised environmental risks are:

- Injurious effects suffered by persons as a result of exposure due to activities of a company, which can be expressed either as an individual risk and/or a societal risk.
- Non-compliance with regulations, which can result in substantial penalties and fines. (Any emission, legal or not, can generate future liabilities.)
- Loss of reputation, business and market shares, if an environmental disaster occurs in the company.

To handle these problems, various models for risk management have been developed. A very simple and general definition of risk management, as implied earlier, is *calculating combined risks and impacts and making decisions*. However, risk management applies to a very wide range of different situations, and it is necessary to describe the context of a risk analysis in a very comprehensive way, including the objectives of the analysis, the framework and goals. The general process of risk management can be subdivided in three main phases as below:

(i) **Risk analysis (or identification):** This involves the identification of hazards to people and the environment, the determination of the probability of these hazards happening and assessment of the seriousness of possible events.

(ii) **Risk assessment:** This means defining the limits or acceptability of the risk, which can be classified as acceptable or in need of reduction. Its main purpose is to define the company's reference safety policy. This policy, based upon the risk assessment, should include risk characterisation schemes, acceptable levels of risk, safety schemes, safety goals, safety standards and regulations. The risk limits for different activities of the company are set using a three-band index, consisting of:

- the maximum acceptable risk limit should never be exceeded,
- the negligible risk limit, indicating the level below which it is not economically or environmentally sensible to try to reduce the risk,
- the gray area between the above risks, within which a trade-off must be made between cost effectiveness of risk reduction and the goal of achieving an acceptable level of risk.

The setting of risk limits in companies is a management procedure. But, it may also be a political choice for a municipality or even a parliament, depending on the consequences of an unexpected event (e.g., the risk of setting up a nuclear power plant).

(iii) **Risk control:** This involves the design and implementation of risk reducing measures and controls. This process includes various actions to be taken to achieve optimal allocation of a company's resources for risk control and reduction. The risk control analysis includes identification of measures to reduce risks, estimation of likely reductions to be achieved by various measures and estimates of the costs of the measures. Possible risk control measures should be ranked and listed to assist in further decision-making. Risk control in a company can be achieved by measures affecting the layout of different production facilities or procedures, application of additional safety devices, and, of course, the choice of less hazardous activities. Further risk reductions may be creating and maintaining safety areas around specific plants or devices.

To complete the risk control analysis, there are some useful tools that can be used such as multi-criteria decision analysis, cost-benefit analysis, sensitivity analysis and cost effectiveness analysis. The best available method for incorporating uncertainties should be used. During the risk control phase, companies should also prepare detailed emergency procedures with the public, and the responsible authorities should be informed. A well-structured plan will play an important role in limiting the effects on people's health and the environment. Figure 8.5 shows the general procedure for risk assessment.

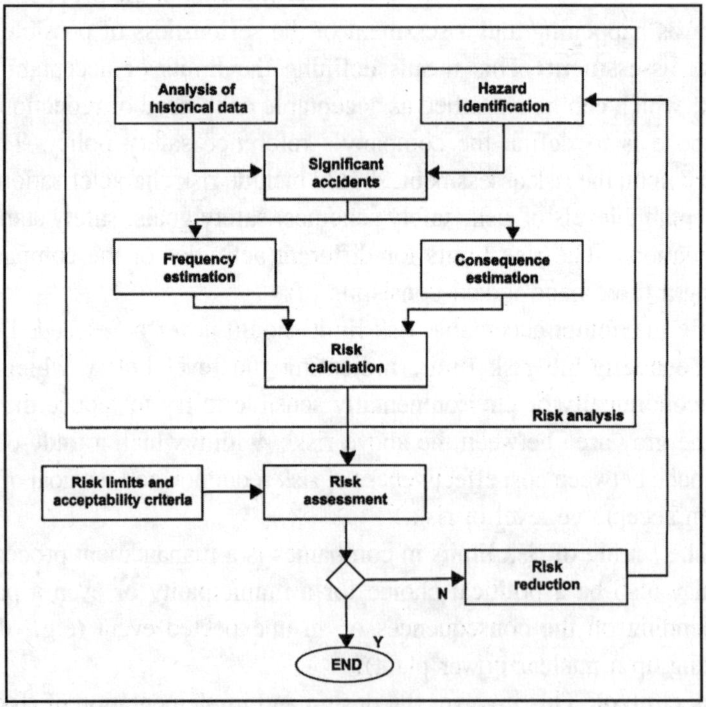

Figure 8.5 A General Procedure for Risk Assessment

LEARNING ACTIVITY 8.4

State three probabilities you would consider in risk analysis of a petroleum product storage tank in a refinery.

Note:

a) Write your answer in the space given below.
b) Check your answer with the one given at the end of this book.

8.7.2 Ecosystem Approach to Risk Assessment

Ecological assessment is the determination of harmful impacts at the population, community, or ecosystem level and may entail field studies, laboratory studies, or both. Measuring or accurately predicting a significant change in an ecosystem may require years or decades of study, yet the risk assessor and risk manager are faced with much more compressed timelines. For example, it is difficult to isolate easily studied areas of the wetlands from the surrounding ecosystem that supports it, which may require the risk assessor to include caveats and large uncertainties in the risk assessment.

Given this situation, most ecological assessments have focussed on measuring structural components of the ecosystem, including the size and make-up of the habitat, biomass and standing crop of important plants and animals, the abundance and diversity of plants and animals and other measures.

Functional Assessment

Functional measurements are useful for understanding the ecological health and sustainability of wetlands. These include hydrologic flux and storage, biological productivity, biogeochemical cycling and storage, decomposition and wildlife habitat. A functional assessment is based on the concept that variables (functions) that integrate key ecosystems—level wetland processes can be used as metric to compare impacts and quantify functional loss when compared to reference wetlands of the same hydro geomorphic classification (Brinson, 1993b; Richardson, 1994). Functional assessment can be a powerful tool for evaluating ecosystem level impacts and risks within a timeframe necessary for risk assessment.

One of the main strengths of this approach is that it causes the risk assessor and risk manager to identify and resolve problems in the context of what a wetland does on the landscape. Although human values may ultimately drive the risk assessment, the inseparable linkage of values to ecological functions, particularly for constructed wetlands, becomes evident as this approach is planned and implemented. Moreover, it provides the risk manager with information on the functions that must be maintained or restored in order to provide the desired outcome.

Risk Estimation

This refers to the estimation of the potential impact or the outcome or consequence of an exposure on a receptor (target) is estimated. In the case of accidental and sudden releases, the assessor will proceed to risk estimation, if the hazard assessment indicates a potentially significant probability or frequency of occurrence of an event and/or a consequence. The types of damage caused by sudden releases include acute toxic effects through dermal contact and inhalation of fumes, burns from fires and explosions and damage to humans, the built environment (e.g., flash floods due to dam failure), etc.

Chronic health effects are, however, unlikely to result from single, brief exposure to accidental releases. Chronic health effects on humans caused by exposure to low level releases of chemicals during day-to-day operations have generally been the focus of risk estimation in the context of EIA. Chemicals can exhibit a wide variety of chronic health effects in humans such as:

- sensitisation (e.g., the induction of dermatitis following regular skin contact with a chemical);
- neurotoxicity, or damage to the central nervous system, causing behavioural changes, tremors, etc.;
- teratogenicity, or damage to the embryo and foetus, leading to birth defects;
- mutagenicity, or damage that results in changes to the DNA structure in genes;
- carcinogenicity, or the development of malignant tumors and neoplasm (i.e., new growth).

The dose-response relationship is fundamental to the interpretation of toxic effects, i.e., there is a direct relationship between the exposure intakes, the duration of intake and the severity of the toxic effect. The basic quantitative difference in risk assessment for carcinogens versus that for non-carcinogens centres on the concept of thresholds. While non-carcinogens have threshold doses below which they fail to induce any discernible adverse health effects, carcinogens are often assumed to have no such effects, and carcinogenic risk is dependent on total dose, independent of whether the dose occurs over a short or long time period.

Carcinogens tend to dominate public health risk when they are present, and hence it is important to differentiate between chemicals that induce cancer and those that do not. Carcinogens can be grouped into two categories: genotoxic (i.e., carcinogens that operate through an initial effect on DNA or chromosomes that initiates cancer) and non-genotoxic (i.e., carcinogens that operate through chronic cell damage). The health risk to individuals is estimated separately for carcinogens and non-carcinogens. For non-genotoxic carcinogens and non-carcinogens, the ratio of the calculated intake to the relevant accepted daily intake (ADI) or tolerable daily intake (TDI) is computed, which is compared with the hazard index (HI).

For genotoxic carcinogens, the US EPA has applied a modelling technique to estimate the largest possible linear slope to extrapolate from high doses typical of experimental studies to the much lower doses typical of environmental exposures. A carcinogenic potency factor, called the potency factor, is computed from the data, which when multiplied by the calculated intake of the carcinogen, produces an estimate of the additional or incremental cancer risk to the individual as a result of exposure via the various exposure pathways. The risk is expressed as an estimated annual or lifetime risk of the chance of developing cancer (e.g., one in one million or $1 : 10^{-6}$).

Table 8.3 gives some examples of impacts amenable to risk assessment:

Table 8.3 Impacts Amenable to Evaluation by Risk Assessment

Type of project	*Project examples*	*Types of impact amenable to risk assessment*
Extractive industry	Mines, quarries, cement manufacture	Damage to vegetation due to dust emission; effluent discharge; transportation risk; potential for flooding; health risks, associated with stack emissions
Waste management	Landfills, incinerators, chemical treatment plants, sewage treatment works	Health risk associated with stack emissions to atmosphere and release of leachate to groundwater; fire and explosive risk from landfill gas; accidental releases due to equipment failure, spillages or operational errors; transportation risk
Transport and infrastructure	Roads, railways, airports, harbours, dams	Structural failure; flood damage; transportation risk and accidents to road users; airborne accidents; quayside accidents
Chemicals, manufacturing, textiles, pulp and paper	Petrochemical plants, chemical and pharmaceutical production, dyeing works, paper and textile manufacture	Effluent discharge; ecological impacts in receiving waters; equipment failure; accidental chemical releases; fire and explosion risks
Forestry and agriculture	Afforestation, drainage and flood defence works, fish farming, livestock-rearing installations	Ecological and habitat changes; water and effluent management

Section 8.8 will introduce you to another essential tool called Rapid Urban Environmental Assessment.

8.8 RAPID URBAN ENVIRONMENTAL ASSESSMENT

Little information is readily available on environmental conditions, the interaction between urban development and ecosystems, or the managerial setting that exists to respond to environmental problems in the cities of the developing world. Recent attempts to develop such information have been incomplete because they:

- focussed on a limited number of variables that do not present a complete picture of key environmental issues;
- took a narrow perspective by examining only one sector within the city;
- required several years of intensive, multidisciplinary research and analysis;
- did not develop a set of urban environmental data.

As a methodology, rapid urban environmental assessment (RUEA) draws its inspiration from rapid rural appraisal and participatory rural appraisal. However, urban assessment is much less anthropological and community-focussed than its rural counterparts primarily because cities involve much larger populations and spatial areas.

Description of Methodology

The RUEA methodology consists of a three-step process as under:

(i) **Measurement:** This involves data collection through an urban environmental data questionnaire.
(ii) **Observation:** This involves preparation of an urban environmental profile, using data from the questionnaire and research assistance from local investigators.
(iii) **Validation and action:** This involves discussion of the results through a series of consultations, culminating in a priority-focussed workshop.

This three-step process is adopted to enable local experts and citizens to rapidly assess the state of the urban environment.

Users of the Tool

Local governments of urban or urbanising localities use RUEA, as it provides a comprehensive diagnostic approach to the determination of a wide range of environmental problems and solutions. Local planning bodies may particularly steer its implementation.

Application/Scope

The process of RUEA (i.e., data collection, profile and consultations) can provide an informational and consensual basis for preparing an urban environmental management strategy with a view to accelerating the improvement of environmental conditions in cities, especially by integrating key aspects of urban policy and environmental management.

The urban environmental management strategy builds on existing sector and project work but emphasises continuity in decision-making to implement agreed policies and approaches. It should provide a decision-making framework for private investments (for public) while recognising that the investments will be primarily private (by households and firms). It, therefore, requires a participatory process among decision-makers in government and the private sector, often using working groups of officials in consultation with technical specialists and key private and informal sector actors, as they agree and commit themselves to act on the policies and strategies they themselves will define.

This strategy can be developed in a number of different ways but should generally include consideration of health effects and environmental damages (i.e., costs), comparison of alternative long-term strategies to achieve environmental quality goals at the lowest economic cost to the urban region, identification of appropriate policies and instruments to implement the least cost strategy and an assessment of its institutional and financial feasibility.

In Section 8.9, we will introduce you to a relatively new concept called eco-mapping. This is extremely useful for small and medium enterprises (SMEs) to manage their activities.

8.9 ECO-MAPPING

Environmental management standards such as ISO 14001, or the EMAS regulation, which are becoming prevalent throughout the world at the moment, threaten the very existence of SMEs.

Several million micro-enterprises and SMEs in many parts of the world grapple with the problems of managing the environmental impact of their activities due to, among others, inadequate tools and resources (*http://www.inem.org/htdocs/inem_tools.html*). The greatest problem they face, however, is one of awareness and changing behaviour. Nevertheless, they will soon be required to show customers, public administrations, non-governmental organisations, insurance companies, neighbours and professional associations that they have implemented environmental management.

What is Eco-mapping (or eco-maps)?

Eco-mapping is a visual and easy-to-use tool which enables employees to get involved in environmental management in a company. The results of a quick and visual environmental review can be equivalent to that of expensive scientific studies conducted by consultants. If the results of a scientific study can be compared to a high-resolution image, those of eco-maps, which do not cost anything, can be compared to the Polaroid photograph of environmental management. Both will enable you to take positive action.

Eco-mapping is a road map of a site, a shop floor, a workshop, etc., which can lead to improved environmental management and which can provide a solid basis for a more formal environmental management system according to ISO or EMAS. In essence, it is

- an inventory of practices and problems;
- a systematic method of conducting an on-site environmental review;

- a collection of information which shows the current situation using pictures;
- a work and awareness-raising tool;
- a do-it-yourself tool for SMEs;
- a tool which allows employee involvement and participation.

Eco-maps can be used to determine:

- a site in its urban environmental context;
- consumption of water and discharge of wastewater;
- storage of flammable, dangerous or hazardous substances in relation to groundwater;
- emission of air pollutants and dust, odour and noise;
- consumption of energy.

How to use Eco-maps?

The materials needed are A4-sized papers and a photocopy machine. Eco-mapping is done at the end of the accounting year. It is updated once a year, or when the site is renovated or the activities are extended. From small manufacturing and service companies to large structures and local authorities can use the maps. Eco-mapping involves:

- **Mapping of the urban situation:** Make a map of the site including car parks, access areas, roads and the surrounding environment. The map should reflect the real situation.
- **Mapping of the site:** Draw the outline of the site using a scale and showing the interior spaces. This map should be copied and will be the basis for the work to be done. The maps should show the real situation and should be simple, recognisable and in proportion. They should have a date, a name and a reference. Certain objects such as machines, boilers, etc., will have to be in order to orient you straightaway. If a site covers very different areas, mapping can be done for different areas separately and then merged.
- **Developing symbols:** Symbols can be developed by using either hatched lines when the problem to be studied is small or circle if the problem is large. The more serious the problem the thicker the circle.

In order to improve the quality of the eco-maps, standardised pictograms can also be used.

So far, we discussed some of the environmental management techniques with particular reference to urban contexts. They are useful in environmental management. But, for strategic reasons, environmental education is imperative. We will discuss this aspect of environmental management, next.

8.10 ENVIRONMENTAL EDUCATION

Environmental education is a learning process that increases people's knowledge and awareness about the environment and associated challenges, develops the necessary skills and expertise to address the challenges, and fosters attitudes, motivations, and commitments to make informed decisions and take responsible action (UNESCO, Tbilisi Declaration, 1978) (*http://www.epa.gov/enviroed/epagoals.html*). It does not advocate a particular viewpoint or course of action.

The components of environmental education are:

- awareness and sensitivity to the environment and environmental challenges;
- knowledge and understanding of the environment and environmental challenges;
- attitudes of concern for the environment and motivation to improve or maintain environmental quality;
- skills to identify and help resolve environmental challenges;
- participation in activities that lead to the resolution of environmental challenges (UNESCO, 1978).

Environmental education is essential for improving the quality of our lives. Let us now discuss this aspect.

Protecting Human Health

The link between environmental challenges and human health is clear. Lead poisoning from paint and pipes, air pollution, pesticides in water and food supplies, increased threats of skin cancer from depletion of the ozone layer and other environmental and health issues are a veritable cause for public concern. Environmental education can help prevent or mitigate environmental human health problems by providing the public with information on the causes of environmental pollution, how pollutants may affect health, how to assess real versus exaggerated risks and how to make informed and responsible decisions that prevent or mitigate the effects of pollution on health.

Advancing Quality Education

Educators and public officials generally believe that improvements are needed in the nation's public education system to enhance student learning. Many educational scholars and practitioners agree that students are not doing well at thinking, reasoning, analysing or problem solving. Many goals of education reformation emphasise the need to strengthen core subjects, improve critical thinking and problem-solving skills and relate learning in the classroom to the needs and issues of the community.

Environmental education has a tremendous potential to contribute to these goals of education reform. For example, it can strengthen teaching in many core subjects, especially science, because science is the basis for solving many of our environmental challenges. At the same time, environmental education can promote interdisciplinary teaching because environmental topics can be addressed from many different perspectives, including scientific, historical and cultural.

Environmental education can also bring local environmental challenges into the classroom to improve analysis and problem-solving skills.

Creating Jobs in the Environmental Field

Protecting the environment creates new jobs. Employment opportunities in the environmental field cover the spectrum of careers from manual labour to high technology and management. For example, there is an increasing demand for individuals with specialised scientific and technical skills to develop more effective pollution prevention and control technologies in areas such as air pollution control, environmental energy sources, hazardous waste management, resource recovery and instrument manufacturing. Environmental education and training can help ensure an adequate supply of well-trained environmental personnel to deal with the nation's increasingly complex environmental challenges. Environmental education's emphasis on critical thinking and problem-solving skills equips learners in dealing with rapidly changing technologies in the workplace.

Promoting Environmental Protection along with Economic Development

Our future depends on the ability to use the natural resources in a sustainable manner. Sustainable development poses two fundamental educational challenges. One is to promote informed decisions that are conducive to sustainability and the other is to teach the benefits of integrating conservation with the need for economic development. Environmental education contributes to sustainable development efforts by demonstrating ways to promote informed decision-making and teach the benefits of linking conservation and economic development. Environmental education research has identified key strategies for developing education programmes that lead to responsible decision-making, and practitioners have developed programme models for incorporating a range of perspectives into the resolution of issues. These educational tools and strategies can be applied to efforts to ensure sustainable development.

Encouraging Stewardship of Natural Resources

Interest in protecting a nation's natural resources, including species and their habitats, arises from the respect most people hold for the nation's past and a belief in its future. This interest also stems from a strong desire to protect and enjoy nature and to pass it on to our children. There is a need to increase the understanding that the health of individual species can be strong indicators of the health of the entire environment and that biological diversity has ecological and economic importance. Environmental education enhances the understanding of the public about the need for healthy plant and animal life and biodiversity. It also educates the public about the impact of their actions on natural ecosystems and how positive steps taken to minimise impacts on these ecosystems will translate into improvements in our overall environment.

SUMMARY

Following our discussion of the major environmental management (EM) tools in Units 4 to 7, in this Unit, we dealt with some of the other EM techniques with particular reference to urban contexts. We began the Unit by discussing environmental monitoring. We then explained the use of mathematical modelling to deal with environmental issues, and in that context, introduced forecasting and growth modelling. We also discussed sensitivity analysis and geographical information system. The other techniques we discussed in the Unit are environmental profile, environmental technology and risk assessments, RUEA and eco-mapping. We closed the Unit by bringing to the fore the importance of environmental education.

REFERENCES

Brinson, M.M., 1993."A hydrogeomorphic classification for wetlands". U.S. Army Corps of Engineers, Vicksburg, MS, USA. Wetlands Research Program Technical Report. WRP-DE-4, Waterways Experiment Station., Washington, DC

Meadows D H, Meadows D L, Randers J, Behrens W W, 1972. *The Limits to Growth III,* Potomac Associates, New York, Universe Books

Meadows, D.H., Meadows, D.L. and Renders, J. 1992. Beyond the Limits, Global Collapse or a sustainable future. Earthscan Publication Limited, London.

Richardson, C.J. 1994. Ecological functions and human values in wetlands: a framework for assessing forestry impacts. Wetlands 14:1–9.

UNESCO, 1978. Tbilisi Declaration: Final Report of the Intergovernmental Conference on Environmental Education, *www.ericdigests.org/2003-2/outdoor.html*

SUGGESTED READINGS

Basset, R. and Tomlinson, P. 1988. Monitoring and auditing of impacts. *In:* Environmental Impact Assessment, Theory and Practice, P. Wither (Editor). Unwind Hyman, London.

Canter, L.W. and Fairchild, D.M. 1986. Post-Eyes Environmental Monitoring. *In:* Methods and experiences in Impact Assessment. H.A. Becker and A.L. Porter (Editors). A special issue of *Impact Assessment Bulletin*, Volume 4, No. 3-4, Spring, 1986. International Association for Impact Assessment.

Chapman, D. (editor). 1992. Water Quality Assessment: A Guide to the Use of Biota, Sediments and Water in Environmental Monitoring. Published on behalf of UNESCO, WHO and UNEP. Chapman & Hall, London.

De Jog, P. 1988. Uncertainty in EIA. *In:* Environmental Impact Assessment, Theory and Practice. P. Wathern (Editor). Unwind Hyman, London.

Meadows, D.H., Meadows, D.L., Renders, J. and Behrens, W.W. 1972. *The Limits of Growth*. Earth Island Ltd., London.

Novella, V.T. and Marshier, M.W. 1993. Risk Assessment Methods: Approaches for assessing health and environmental risks. Plenum Press, New York.

Rosenthal, J.C. 1980. *Introduction to Population Modelling*. Birdhouse, Boston.

World Health Organisation. 1989. Environmental Impact Assessment: An Assessment of Methodological and Substantive Issues Affecting Human Health considerations. University of London, WHO Report No. 41.

World Health Organization. 1991. Environmental Health Criteria Series Catalogue, 1991. United Nations Environment Program, International Labor Organisation and WHO, Geneva.

UNIT 9

Environmental Design

STRUCTURE

OVERVIEW

In the last two decades, the manufacturing sector has been undergoing changes to meet diverse requirements that include maintaining high quality at low cost, staying competitive in a global marketplace, and meeting consumer preferences and regulatory demands for reduced environmental impacts. As a result, environmental management is increasingly becoming proactive i.e., anticipating and preventing environmental problems. Designing for the environment or environmental design also contributes to proactivity. It is a strategy for effectively organising environmental issues. In this Unit, we will discuss the tools available for integrating environmental concerns in the process of designing industrial products, buildings and developmental planning. We will begin the Unit by introducing the concept and principles of environmental design (ED). We will then discuss ED for manufactured products and construction. We will also discuss ED in the context of developmental planning.

LEARNING OBJECTIVES

After completing this Unit, you should be able to:

- explain the basic principles, objectives and potential of environmental design (ED);
- discuss environmental design considerations in the context of manufactured products, construction and developmental planning.

9.1 ENVIRONMENTAL DESIGN (ED): AN EXPOSITION

Environmental design (ED) or designing for environment is a perspective in which environmental aspects of an existing or new product, process or facility, design/redesign are optimised systematically and continuously for minimal adverse impact on the environment. In other words, it helps in designing and developing environmentally benign products and processes that have the lowest life cycle environmental impact. Acting as a catalyst, ED helps provide businesses with information to make environmentally informed choices. Note that this approach is also referred to as life cycle design, eco-design, sustainable design or design for eco-efficiency.

9.1.1 Principles of ED

In order to provide a context for the discussion of the principles of environmental design (ED), we will first look into its basics. ED is a design of products to meet environmental as well as utilitarian objectives. The core environmental objective is sustainability of a product and process. As the environmental management tools evolved, it became evident that environmental benefits can actually be built into the plans and design, and in so doing, products and processes can contribute to the improvement of the environment.

To elaborate, the environmental issues relating to processes and products are addressed, generally, after their implementation, and this involves redesigning of the processes and products to the extent possible. However, such tasks are expensive and time consuming and rarely attain the required level of environmental improvements. It is, therefore, necessary to evaluate potential environmental impacts at the designing stage, and it is this principle that underlies the concept of environmental design. In other words, there is a need to shift the conventional focus of processes and products from utilitarian values to environmental values. This shift in focus results in processes and products that are environmentally sound as well as utilitarian oriented. The concept of ED is thus useful in the manufacturing of products and in the processes of developmental planning.

The genesis of many environmental problems is high population levels, rapid population growth and unequal population distribution. The natural systems are getting overloaded and are unable to provide resources to meet the basic minimum needs of the ever-growing population. The result is the vicious cycle of poverty, rapid population growth, environmental deterioration and more poverty. Contributing to this vicious cycle are the results of economic activities that were once perfectly acceptable and even desirable but are now considered unacceptable because of their impacts on the natural resource base. Till recently, humanity was concerned about the ill effects of development on the environment. But now, it is equally concerned about the ways in which environmental degradation can dampen or reverse economic development.

The WCED report of 1987, *Our Common Future*, emphasised that the size, density, movement and growth rate of population cannot be influenced, if population control efforts are being overwhelmed by adverse patterns of development. Since then, the concept of carrying capacity of an ecosystem is being looked upon as a tool for enabling environmentally sound developmental planning. This is born out of the realisation that the preservation of ecosystems and of the productive potential of nature is a fundamental objective of all planning.

Against this backdrop, let us now discuss the key principles of ED below:

(i) The design of a project, process or product must incorporate environmental objectives, in addition to other objectives such as economic efficiency and practicality.

(ii) The core environmental objective is the overall sustainability of:
- the use of resources, including energy;
- the process used in the manufacturing of a product or constructing a project;
- the product or project in operation;
- the product or project at the end of its use or operation, i.e., the actions involved in discarding, recycling or dismantling the product or project.

(iii) While sustainability is the principal environmental concern, other environmental objectives that should be considered are:
- the restoration and enhancement of landscapes;
- the dignity, comfort and safety of workers in workplaces;

- the restoration and enhancement of ecosystem;
- fostering of human creativity and art.

(iv) Environmental objectives that refer specifically to a project, process or product should be stated explicitly, to a level of detail equivalent to those of the other objectives. These objectives are to be agreed upon at the outset by administrators, sponsors and designers.

(v) Since the environment is multi-sectoral and multi-disciplinary, design for environmental objectives is inevitably a multi-discipline exercise and multi-disciplinary activity must be considered within the environmental design.

(vi) The design of a project, process or product should proceed from its initiation by the conscious definition and integration of environmental and other practical considerations. This requires the consideration and statement of design process that is product-or project-specific.

LEARNING ACTIVITY 9.1

List the core objectives of environmental design.

Note:

a) Write your answer in the space given below.
b) Check your answer with the one given at the end of this book.

9.1.2 Benefits of ED

ED offers businesses the opportunity to enhance environmental performance, while simultaneously improving their bottom line (*http://www.nrc.ca/DFE/ehome/ehome.html*). Companies that apply the concept of ED, for example, find that it:

- reduces environmental impact of products/processes;
- optimises raw material consumption and energy use;
- improves waste management/pollution prevention systems;
- encourages good design and drives innovation;
- cuts costs;
- meets user needs/wants by exceeding current expectations for price, performance and quality;
- increases product marketability.

ED can also provide a means for establishing a long-term strategic vision of a company's future products and operations. In essence, as an enabling force, ED helps shape sustainable patterns of production and consumption. In this context, consider the following:

(i) **Increased innovation:** By incorporating ED into product design/ development, companies gain a fresh perspective on established practices, resulting in new ideas and solutions (e.g., new product/service concepts, alternate production techniques, increased employee participation and greater creativity).

(ii) **Greater ability to compete, add value and attract customers:** There is a growing global demand for environmental quality in products and services. Incorporating ED into product design can help companies:
- meet emerging market demands;
- differentiate their products in the marketplace;
- improve their image and win customer attention;
- attract investment.

(iii) **Become more cost-effective:** ED targets opportunities for cost-reduction at all stages of a product's life and ensures the greatest reduction in environmental effects/releases per rupee invested. This results in:
- reduced production costs;
- increased product quality;
- elimination of compliance costs;
- increased return on environmental investments.

(iv) **Reduce environmental impacts and liability:** By decreasing a product's impact on the environment, ED helps companies:
- ensure compliance with environmental regulations;
- reduce uncertainty with respect to future environmental requirements;
- improve access to insurance and financing;
- achieve better community relations;
- contribute to a better local, regional and global environment.

(v) **Gain a systems perspective:** ED, which focusses on a product's life-cycle, helps companies create corporate links between product design, supply chain management and sales/marketing, thereby providing:
- an overall, systemic view of company operations;
- a mechanism for cross-functional teams to continuously improve products.

The three sample success stories listed below should attest to the benefits of ED:

Example 1

Interface Flooring Systems (IFS), a leading global supplier of floor coverings, fabrics, chemicals and interior architectural components, has been using product redesign and material formulation to eliminate targeted toxic elements and compounds in its production processes. Through its efforts, IFS:

- eliminated costs of toxic substances;
- improved product quality and performance;
- reduced unsaleable materials from 5.4 to 0.4% since elimination of toxic substances was a factor in improving product quality.

IFS saved over 90 million USD from 1995 to 1999, as a result of its in-house, zero-waste initiative (Rowledge, Barton, and Brady, 1999)

Example 2

Electrolux, the Swedish global appliance manufacturer, used a combination of design and life-cycle tools to improve energy and water efficiency in its product lines. As a result:

- It expanded its market share in professional refrigeration equipment from 5% in 1997 to 14% in 1998 with its Explorer line, which was in the top range of energy-efficient appliances.
- Its Master System for apartment washing machines optimised the use of electricity, water and detergents, reducing laundry costs in apartment buildings by 50 percent (Electrolux, Environmental Report, 1998).

Example 3

IBM Sweden designed glassware made from recycled cathode ray tubes (CRTs). The face of the CRT comprises two thirds of the weight of the glass and does not contain harmful lead. The result is that the majority of glass can be recycled into glassware. The other parts, containing toxic materials, can be recycled into CRTs at IBM Holland in the Netherlands. The results are cost-savings in recycled materials and greater awareness among marketing staff regarding the value of recycled materials (Belmane, 1999).

The above discussion clearly illustrates the advantages that result in adopting DFE in companies. Let us discuss the key concept of DFE in Subsection 9.1.3, i.e., external and internal drivers.

9.1.3 Motivation for ED

Motivation to implement ED can come from either or both of the following:

(i) Within the company itself, referred to as *internal drives*.
(ii) Within the immediate surroundings, referred to as external drives.

Let us now discuss both of these drivers.

Internal Drivers

The internal driving forces in ED should consist of:

(i) **Need for increased product quality:** A high level of environmental quality will raise product quality in terms of functionality, reliability in operation, durability and reparability.

(ii) **Image improvement:** Communicating a product's environmental quality to users through an environmental seal of quality such as the Environmental Choice Label or a good report in consumer tests, can improve a company's image significantly.

(iii) **Need to reduce costs:** Companies can combine ED strategies with financial benefits by:
- purchasing fewer materials for each of its products;
- using energy and auxiliary materials more efficiently during production;
- generating less waste and lowering disposal costs;
- disposing of hazardous waste.

(iv) **Need to stimulate innovation:** ED can lead to radical changes at the product system level—the combination of product, market and technology. Such innovations can provide entry into new markets.

(v) **Employee motivation:** Employee morale, generally, increases when employees are empowered to help reduce the environmental impact of the company's products and processes. ED can also boost employee motivation by improving occupational health and safety.

(vi) **Sense of responsibility:** A growing awareness that business must play an important role in working towards sustainable development can provide a strong incentive for implementing ED.

External Drivers

The external driving forces in ED should consist of:

(i) **Government policies:** Product-oriented environmental policy is growing rapidly in northern Europe, the United States and Japan. Some trends along these lines are given below:
- Legislation on extended producer responsibility and take-back obligation (e.g., Germany has introduced a take-back obligation for goods such as television sets, computers and cars. The U.S. Environmental Protection Agency requires discharge disclosures for certain types of generators).
- Introduction of eco-labelling programmes for products or product groups.
- Requirement to provide environmental information on products and processes, requiring businesses to pursue more proactive environmental communication policies.
- Development of industrial subsidy programmes to stimulate ED activities and encourage companies to carry out research into potential environmental improvements.
- Termination of subsidies on energy-intensive production methods and energy/raw material consumption.

(ii) **Market demand/competition:** The needs/wants of suppliers, distributors and end-users are powerful drivers for environmental improvement. The trends include the following:
- Requirements by some companies—generally large corporations—for environmental safeguarding declarations from suppliers. Some companies are systematically looking at their entire supply chain and imposing the new environmental standards or other measures of environmental performance.
- Boycotts or other actions by consumer organisations/environmental groups. For example, Green Peace successfully pressured industry to develop *Green Freeze*, an ecologically efficient refrigerant made of propane and butane that can replace environmentally harmful chlorofluorocarbons (CFCs).
- Environmental requirements incorporated into consumer product testing. If a product fails to get a high score on these requirements, it will no longer qualify for the title of best buy or good choice, no matter what

other excellent features it may possess. Good environmental ratings can increase market share.

- Increased implementation of "responsible care programmes" in many industries, resulting in more companies with experience in cleaner production. In cases where intense competition exists for a particular product, companies with a good environmental profile can have an edge.

(iii) **Trade/industrial organisations:** These organisations often encourage member companies to take action on environmental improvement and/or may impose penalties on companies that do not take the required action. As well, standardisation organisations are expanding all existing norms and standards to include environmental issues. The ISO 14000 series will become the international standard for certifying environmental management systems. It is expected that product-related aspects such as the obligation to collect and publish environmental data, will be incorporated in this standard.

(iv) **Waste charges:** Waste-processing charges such as land-fill and incineration costs are likely to increase, based on the principle of polluter pays. The prevention of waste and emissions, re-use and recycling will consequently become more economic.

(v) **Environmental requirements for design awards:** Several respected design competitions have now stipulated that contestants must provide specific environmental information on their products. One example is the German Industry Forum (IF) Design Award, affiliated with the Hannover Messe, which has a five-year plan to obtain environmental information on aspects such as packaging materials used, reusability and warranties. Other international design competitions now proactive with regard to the environment are as follows:

- IDEA award in the United States.
- G-Mark award in Japan.
- Form Finlandia award by Nestle.
- Excellent Swedish Form by the Swedish Design Council.
- Brown Competition in Germany.
- ION award in the Netherlands.

9.1.4 ED and Other Environmental Practices

ED is designed to help companies adopt environmental practices that will lead to a more sustainable and healthier society. It supports as well as works within the context of other environmental initiatives, some of which we will touch upon, next.

Sustainable Development

One of the basic assumptions underlying sustainable development is that environmental considerations must be entrenched in economic decision-making. Sustainable development initiatives are increasingly widespread among individuals, communities, industry and governments around the world.

Industrial Ecology

This term encompasses the practices of scientists, engineers and manufacturers to achieve more sustainable industrial production and consumption for local, regional and international economies by:

- examining the environmental costs of industrial production/consumption patterns;
- addressing the effects of invisible and persistent toxic chemicals on the earth's ecological systems.

"In a sustainable society, durability and recycling will replace planned obsolescence as the economy's organising principle, and virgin materials will be seen not as a primary source of material but as a supplement to the existing stock" (Brown and Shaw,1982).

Pollution Prevention (PP)

PP focusses on process and product improvements in order to avoid environmental problems before they occur. It is economically and environmentally superior to traditional end-of-pipe controls or clean-up strategies. ED dovetails with PP by focussing on products and processes involved specifically in manufacturing. While many ED strategies incorporate PP, it goes beyond PP practices by also examining product functionality and services. PP, during the manufacturing process, saves costs with regard to disposal, raw materials/consumables, ventilation equipment, maintenance-ducts, motors, internal balancing operations, pollution prevention equipment, health-workers protective equipment and training, regulatory compliance-approval from governments, etc.

A number of measures can be taken to prevent pollution during manufacturing. These include controlling pollution at source (e.g., substitute materials and change form of material to reduce emissions), enclosing the process (e.g., prevent release, accomplished with sealed vessels and piping), suppressing emissions (e.g., water sprays—dusty processes or liquids and gaseous—gas blanket) and changing the process entirely (e.g., degreasing from chlorine-based to high-pressure steam and soldering from traditional acid etching/fluxes/lead to different base materials, VOC-free fluxes, lead-free solder.)

Environmental Management Systems (EMS)

EMS such as ISO 14001 is an organisational approach to facilitate environmental evaluation and management. Except in cases of legal compliance, an EMS does not set or demand specific levels of performance in relation to product or process design. The core requirement for EMS is that organisations have a reasonable amount of information on the environmental effects of their products and processes and, in turn, seek continuous improvement. Pollution prevention (PP) is typically a part of EMS. It is estimated that INR (Indian Rupee) 1 spent on prevention saves INR 10 otherwise spent at end-of-pipe or INR 100 expended on environmental remediation.

ED is complementary to EMS. It augments the organisational approach by including product-oriented environmental evaluations and improvements. Manufacturers using ED strategies take into account the environmental aspects of a product's use and end of life, and apply this information during its design, production and distribution.

The benefits of EMS include:

- Cost savings from greater efficiency in processes, waste reduction, materials and energy use.
- Increased ability to meet customer/supplier requirements.
- Greater competitive advantage.
- Regulatory compliance and reduced liability.
- Improved community relations.
- Greater company appeal for investors.
- Increased employee pride and morale.

Occupational Health and Safety (OH&S)

The ED strategies provide support for programmes related to worker safety during production and worker health in terms of material selection and use. ED helps reduce the need for in-plant emission controls, worker contact with physical or chemical hazards and need for protective equipment.

Improving the performance of health and safety will reduce lost time due to injury and illness, reduce insurance fees and liabilities, improve employee morale and increase productivity.

Table 9.1 illustrates the link between environmental design and OH&S elements.

Table 9.1 The ED-OH&S Link

ED Strategy	*OH&S Elements*
Strategy 1: Physical Optimisation Integrate Product Functions. Easy Maintenance and Repair. Modular Product Structure.	Ergonomics. Material handling. Safety.
Strategy 2: Optimise Material Use Cleaner materials. Reduce material usage.	Material substitution. Hazardous materials.
Strategy 3: Optimise Production Alternate production techniques. Fewer production steps. Less production waste. Fewer/cleaner production consumables.	Process substitution. Chemical substitution. Industrial housekeeping. Hazardous materials. Safety equipment. Ergonomics.
Strategy 4: Optimise Distribution Less/cleaner/re-usable packaging. Energy-efficient logistics.	Material substitution. Material handling/storage.

Source: Canadian Centre for Occupational Health and Safety, Hamilton, Ontario.

We will next discuss the applications of ED for manufactured products.

9.2 ED FOR MANUFACTURED PRODUCTS

Industrial production is responsible for a great deal of the adverse environmental impact that we experience in the world today. A product may cause adverse environmental impacts during its production, use and beyond. For example, in many instances, the most severe environmental disturbances may be due to the impacts of different means of transportation. The disposal of the product after its use may also be one of the reasons for serious soil and groundwater pollution (e.g., landfilling) or air pollution (e.g., incineration). In addition, for many products, severe physical disturbances and pollution result from the extraction of raw materials. All of these environmental problems are largely decided (consciously or unconsciously) during the design of the product.

This points to the fact that the development of a new industrial product involves many decision-making situations where alternative solutions to a given design problem must be ranked, and priorities given, according to relevant criteria. Many different and very diverse criteria may be applied. Among the important demands on the new product may be:

- **Performance:** The product should actually meet the need that has been recognised.

- **Appearance:** The company may wish its products to have a certain common appearance or the product may belong to a series of products with a certain common look.
- **Manufacturing costs:** The manufacturing costs are essential for determining the later consumer price of the product.
- **Durability:** The company must seek some compromise between the customer demand for high durability and its own wish to be able to market and sell new products frequently.
- **Safety:** The product should be safe to use, thus reducing the company's liability due to accidents.
- **Service:** The product should be easy to service and maintain.
- **Ease of manufacture:** The design of the product must consider the constraints of the manufacturing facilities and the capabilities of the company.

The environmental design of industrial product implies that the traditional product development criteria listed above must be extended to include criteria concerning the environmental impacts of the products throughout its entire life-cycle, and therefore, must be based on a life-cycle assessment of the product. Let us now discuss the importance of the stages of a product life-cycle in the design process in Subsection 9.2.1.

9.2.1 ED Considerations in Product Life Stages

The implications of the different life-cycle stages that need to be considered in the design process are many and varied. Consider the following:

- **Extraction of raw materials:** Because the extraction of raw materials often takes place in scarcely populated regions remote from residential areas, there tends to be a slack concern about the environmental consequences. However, the impacts due to physical disturbances (e.g., destruction of habitats, extinction of species) and emissions (e.g., local pollution) may be severe for activities such as mining of minerals and extraction of mineral oil. Indeed, for many products, the most serious environmental impact may occur in the extraction of raw materials. These environmental effects should be considered as a consequence of the decisions made during the product design process when different materials are considered for the product.
- **Manufacture:** Product developers decide what materials and processes should be used for the manufacture of the new product. They influence the environmentally relevant emissions, physical disturbances and output of solid waste and wastewater that occur during production. Apart from the external

environment, their choice of processes and materials also influences the working environment in the production plant.

- **Distribution/transportation:** The weight of the product is decisive for energy consumption and emissions due to transportation in the distribution of the product. For some consumer products, the use of packaging material is a substantial part of the total resource consumption and the disposal of the used materials causes some of the most serious environmental loadings of the whole product life-cycle. The packaging of the product will often be influenced by the product design. An example of serious environmental problems due to transportation early in the life-cycle (say, namely distribution of the raw materials) is the regularly recurring disasters involving marine and coastal oil pollution from wrecked oil tankers. This is an inevitable consequence of the use of mineral oil as a resource, be it for the supply of heating or electricity or for the production of polymers. It should, therefore, be considered a consequence of the selection of those types of materials in the product design process.
- **Use and maintenance:** For most electrical equipment, the use stage gives rise to the most serious environmental impacts through the emissions occurring during the production of electricity. The same will generally hold true for products incorporating combustion engines (e.g., vehicles, various machine tools, etc.). The product design team must, therefore, focus on the basic reasons for the energy consumption of the product and aim for optimisation of its energy efficiency. Other possible emissions from the use stage that need attention in design are noise or wastewater. If the use of the product entails resource consumption, this will also be of interest in the environmental product design.
- **Disposal:** Once the life of a product has ended, it must be disposed off. From an environmental and resource point of view, the best method of disposal will generally be to reuse or recycle the product or parts thereof. These methods save resources and reduce environmental impacts due to the extraction of materials. Furthermore, they reduce environmental loadings associated with landfilling (i.e., a major source of soil and groundwater pollution and the physical disturbances), incineration of wastes (e.g., serious air pollution and the combustion slag) and composting of applicable to biodegradable materials. The environmental design should, therefore, be one for disposal and for reuse or recycling. This means that consideration of the potentials for recycling and the possible effects of different forms of disposal should influence selection of materials. Furthermore, the construction of the product should facilitate its dismantling into reusable and recyclable parts.

LEARNING ACTIVITY 9.2

State some of the environmental considerations in the design of an automatic washing machine.

Note:

a) Write your answer in the space given below.

b) Check your answer with the one given at the end of this book.

9.2.2 ED Tools for Products

The decisions made early in the design process pertain to fundamental issues such as the technology concept behind the product and the need it satisfies. That is to say, such early decisions fundamentally determine how a product is to be constructed. Subsequent decisions, i.e., decisions made at later stages when environmental consequences can be predicted with some precision over the product life-cycle will often be restricted to minor details such as the choice between closely related materials or processes.

Designing for the environment will, in most cases, involve one or more of the following strategies:

- Product life extension through better reliability, increased durability, improved serviceability and multi-functionality.
- Material life extension by making the materials and components reusable or recyclable, making the separation of different materials quick and

straightforward, labelling the components and materials to ease separation and collection and keeping potential contaminants, e.g., adhesives, away from recyclable materials.

- Reduced use of materials by making products lighter or smaller while performing the same function.
- Switch from non-renewable to renewable resources either in materials or in energy sources.
- Energy efficiency in use as well as in production.
- Pollution minimisation at all stages of the life-cycle.

The tools available for environmental design of manufactured products include the following:

- **Checklists:** The simplest tools in environmental design are checklists. Generally, checklists specify materials or substances to be avoided, and such checklists are referred to as negative checklists. For example, these checklists may indicate to us to eliminate CFCs, avoid mercury and cadmium, use lead-free pigments and so on. These may be extended to include reduction or minimisation requirements, e.g. for VOC's. Checklists can be developed in a number of ways. One approach is to develop one or more checklists that reflect the market needs (e.g., the requirements of customers and regulators in the countries where you sell). Another approach is to rely on standard lists of environmentally harmful substances.
- **Life-cycle assessment (LCA):** Many of the tools available for environmentally beneficial product development or design have their origins in the process of life-cycle assessment (LCA), which we discussed in detail in Unit 6. As you know, LCA is a technique for estimating and assessing all of the environmental impacts (positive and negative) associated with a product or process from the extraction and acquisition of raw materials, through product manufacture, distribution and use to disposal or recycling. Carrying out a complete LCA on a particular product system is a time-consuming exercise, which requires a large amount of data and involves a certain amount of subjective judgement. It can also be very expensive. On the other hand, the data-gathering stage known as life-cycle inventory (LCI) often prompts many improvements on its own, without recourse to the more contentious impact assessment that follows. Detailed LCA and LCI studies have been carried out on a number of products for comparative purposes.
- **Matrices:** Neither comprehensive LCA nor compilation of an LCI is a practical option for most firms when developing a new product or service, and wanting to improve their environmental performance at the same time. A structured

approach to design or development, which considers the environmental impacts arising from each stage of a product's life-cycle is, however, possible for businesses of all sizes and within projects of all timescales. A simple, qualitative overview of the life-cycle can help to identify which of these approaches is likely to yield benefits. Matrix methods provide one such approach to assessing environmental impacts across the life-cycle. In one simple version, i.e., materials, energy and toxicity (MET) matrix, issues pertaining to materials, energy and toxicity can be identified for three stages of the life-cycle, i.e., production, use and disposal. Table 9.2 illustrates the entries that could be made in a MET matrix.

Table 9.2 MET Matrix: An Example

	Materials	*Energy*	*Toxicity*
Production	Plastics, metal for wiring and element.	Electricity, gas and fuel oil.	Pollution from metals extraction and refining. Pollution from oil extraction, petrochemical plants and energy sources.
Use	Water	Electricity	Pollution from power generation and distribution (SO_2, NO_2, CO_2, etc.).
Disposal		Diesel (fuel for transport).	Possible pollution associated with landfills in general.

- **Eco-indices:** Having selected the ED approach for reduced environmental impact, it is necessary to identify performance indicators to monitor the progress. Examples of indicators that can be useful include percentage of recyclable content, energy consumption per year of typical use, weight of specified materials per product unit, etc.

Environmental Design Examples

Environmental design tools are under development in various countries throughout the world. In many cases, parts of a product are changed for a new version with a better overall environmental performance. However, projects evaluating and redesigning entire products according to environmental criteria are still rare. Consider now the following ED examples:

- The Dutch eco-design project involves a scheme of regular visits by a consultation team with a designer and an environmental specialist to product design groups of companies. The consultation continues throughout the design period with a view to developing environmentally friendly products and changing the environmental attitude of the company. Among the products developed through this project are gas cookers, office chairs and trays for the transport and display of pot plants. The Dutch Ministry of Economic Affairs and the Ministry of Housing, Physical Planning and the Environment sponsor the project.

- The Royal Philips Electronics N.V. have identified five focal areas for eco-designing products, and these are *mass reduction means* (i.e., less use of natural resources, less use of processing chemicals, less pollution during processing, less waste (including EOL), less packaging, less transportation pollution, etc.), *reduction/elimination of hazardous substances* (i.e., less health effects during the manufacturing, use and disposal of products), *energy efficiency means* (i.e., less greenhouse gas emission, less use of battery and less waste), *recyclables means* (i.e., extended material use, extended component/part use, reduction in depletion of resources, reduction in EOL waste and reduction in pollution) and *packaging mass reduction results* (i.e., reduced use of resources, reduced transportation pollution and reduced waste generation). Some of the eco-designed products of Philips are *breakthrough in small signal* (BISS) transistor with reduction in mass (91%), hazardous substances (88%), energy (30-55%) and packaging (91%), *free power radio AE 1000* with reduction in mass (61%), hazardous substances (no cadmium), energy (50%) and packaging (60%) and *S2 and S10 starter* with reduction in hazardous substance (no tritium and lead).
- Countries such as Germany, England, Australia and Taiwan are preparing product take-back regulations. Many companies have already introduced take-back programmes.
- Xerox reuses up to 75% of the components and recycles up to 98% of materials from take-back products. Those components that meet criteria for new components are used again in the Xerox Eco-Series Copiers. The take-back programmes saved $50 million in its first year of operation.
- As of 1997, IBM has saved over $70 million through machine parts reuse and over $7 million through use of recycled commodities.
- Hewlett Packard created a worldwide Equipment Management and Remarketing Division to re-manufacture used products including PCs, printers and scanners. As a result, the company has saved millions of dollars in parts manufacturing, has improved its image as environmentally sensitive, and has gained a greater competitive edge in the global marketplace.

A large number of initiatives have been taken in India for the promotion of environmental design. An overview of these initiatives is presented in Case 9.1. An Environmental Information Centre has been established in India for the promotion of environmental design of products (see Case 9.2).

In Subsection 9.2.3, let us introduce you to a new concept under environmental design, viz., eco-labelling, which is receiving worldwide recognition and many countries have designed their own eco-label as a sign of their responsibility to the conservation of environment.

9.2.3 Concept of Eco-labelling

Eco-labelling is a voluntary method of environmental performance certification and labelling that is practised around the world (*http://www.gen.gr.jp.eco.html*). An eco-label is a label, which identifies the overall environmental preference of a product or service within a specific product/service category based on life-cycle considerations. In contrast to *green* symbols or claim statements developed by manufacturers and service providers, an eco-label is awarded by an impartial third-party in relation to certain products or services that are independently determined to meet environmental leadership criteria. There are many different voluntary (and mandatory) environmental performance labels and declarations. The International Organization for Standardisation (ISO) has identified the following three broad types of voluntary labels, with eco-labelling fitting under the Type I designation:

(i) **Type I:** This represents a voluntary, multiple-criteria based, third party programme that awards a license that authorises the use of environmental labels on products indicating overall environmental preferability of a product within a particular product category based on life-cycle considerations.

(ii) **Type II:** This represents informative environmental self-declaration claims.

(iii) **Type III:** This represents voluntary programmes that provide quantified environmental data of a product, under pre-determined categories of parameters set by a qualified third party and based on life-cycle assessment and verified by that or another qualified third party.

The roots of eco-labelling can be found in the growing global concern for environmental protection on the part of governments, businesses and the public. As businesses have come to recognise that environmental concerns may be translated into a market advantage for certain products and services, various environmental declarations/claims/labels have emerged on products and with respect to services in the marketplace (e.g. natural, recyclable, eco-friendly, low energy, recycled content, etc.). These efforts have attracted consumers (i.e., those who are looking for ways to reduce adverse environmental impacts through their purchasing choices) as well as left them sceptical.

Without guiding standards and independent third party investigations, consumers may not be certain about the claims of the companies about eco-labels.

This concern with credibility and impartiality has led to the formation of both private and public organisations providing third party labelling. In many instances, such labelling has taken the form of eco-label awarded to products approved by an eco-labelling programme operated at a national or regional (i.e., multi-countries) level.

Some examples of eco-labels:

9.3 ED FOR BUILDINGS

Environmental design of buildings involves adopting techniques that ensure that the structure is designed, built, renovated, operated or reused in an ecological and resource efficient manner. Environmental building design is called *green building* design.

A green building may cost more upfront, but saves through lower operating costs over the building. The green building approach applies a project life-cycle cost analysis for determining the appropriate upfront expenditure. This analytical method calculates costs over the useful life of the asset. These cost savings can only be fully realised when they are incorporated at the project's conceptual design phase with the assistance of an integrated team of professionals.

In Subsections 9.3.1 to 9.3.3, we will touch upon some aspects of green buildings.

9.3.1 Green Buildings

Traditional building practices often overlook the interrelationships between a building, its components, its surroundings and its occupants. Typical buildings consume more of our resources than necessary, impact the environment negatively and generate a large amount of wastes. It was observed that a standard wood-framed home consumes over one acre of forest and the wastes created during its construction averages from 3 to 7 tonnes. Often, these buildings are costly to operate in terms of energy and water consumption. And they can result in poor indoor air quality, which can lead to health problems.

They are also a major source of the pollution that causes urban air quality problems and the pollutants that contribute to climate change. For example, they account for 49% of sulphur dioxide emissions, 25% of nitrous oxide emissions and 10% of particulate emissions, all of which damage urban air quality. Buildings, in general, produce 35% of the country's carbon dioxide emissions—the chief pollutant blamed for climate change.

In short, buildings have a tremendous impact on the environment—both during their construction and through their operation. The term *green building* refers to buildings that make use of land, design and construction strategies in such a way as to reduce environmental impacts.

Let us now consider an example.

In the USA, there are more than 76 million residential buildings and nearly five million commercial buildings. These buildings together use one-third of all the energy consumed in the USA and two-thirds of all electricity. By the year 2010, it is estimated another 38 million buildings will be constructed. The challenge is to build them eco-smart in that they use a minimum of non-renewable energy, produce a minimum of pollution and cost a minimum of energy dollars, while increasing the level of comfort, health, and safety of the people who live and work in them. There are many opportunities to make buildings cleaner. For example, if only 10% of homes in the USA used solar water-heating systems, we would avoid about 8.4 million metric tonnes of carbon emissions each year.

What this means is that green building practices offer an opportunity to create environmentally sound and resource efficient buildings by using an integrated approach to design. Green buildings promote resource conservation, including energy efficiency, renewable energy, and water conservation features; consider environmental impacts and waste minimisation; create a healthy and comfortable environment; reduce operation and maintenance costs and address issues such as historical preservation, access to public transportation and other community infrastructure systems. Green buildings consider the entire life-cycle of the building and its components as well as its economic and environmental impact and performance (*http://www.sustainable.doe.gov/buildings/gbprinc.shtml*).

What are the principles to be adopted when designing green buildings? We will take up this question in Subsection 9.3.2.

9.3.2 Principles of Green Building

The principles of a green building include the following (*http://www.sustainable.doe.gov*):

(i) **Energy efficiency and renewable energy resources:** Commercially available, cost-effective energy technologies could reduce the overall energy consumption, for example, in the United States by as much as one-third—worth some $343 billion. Strategies to achieve this include proper siting and airtight construction, as well as installing energy-efficient equipment and appliances and renewable energy systems.

(ii) **Environmental impact:** The built environment has a tremendous impact on the environment. However, your building can interact more positively with the environment, if you pay special attention to preserving the site's integrity and natural characteristics, appropriate landscaping and the selection of materials that have lower embodied energy and those that are produced locally.

(iii) **Resource conservation:** Conserving resources is a cornerstone of green building techniques. There are many ways to conserve resources during the building process. For example, selecting materials that have at least some recycled content can conserve natural resources and virgin materials. Minimising construction wastes can ease the impact on landfills and resources. Installing water-and energy-efficient products can conserve resources, while reducing operating costs. Choosing a green (plant-covered) roof can reduce energy use, cool urban heat islands and prevent storm water runoff as well as contributes to wildlife habitat and air quality.

(iv) **Indoor air quality:** Energy-efficient buildings are more airtight and therefore hold greater potential for indoor air quality problems, especially if not properly ventilated. Building products can contribute to poor air quality. But, by selecting the materials lower in chemicals and toxins and installing mechanical ventilation systems to ensure an adequate fresh air supply, we can reduce adverse environmental impacts.

(v) **Community issues:** Placing green building projects within easy access of public transportation, medical facilities, shopping areas and recreational facilities decreases the need for automobiles and encourages bicycling and walking. In addition, successful green buildings blend into the community, preserving natural and historical characteristics and will utilise existing infrastructure in order to reduce sprawl. Co-housing represents one approach to creating a community of green buildings.

9.3.3 ED Strategies for Building Construction

Environmental considerations in building construction include selection of site, energy use, water use and selection of material for construction. By incorporating the following strategies in building planning and construction, we can make it environment friendly:

(i) **Building siting:** ED considerations in siting of a building include selecting a site well suited to take advantage of mass transit, protecting and retaining existing landscaping and natural features, selecting plants that have low water and pesticide needs, and generating minimum plant trimmings and using composts and mulches to save water and time.

(ii) **Energy efficiency:** ED considerations in energy efficiency improvement of building include the following:

- Select passive design strategies (i.e., shape and orientation, passive solar design and the use of natural lighting can dramatically affect building energy performance).
- Develop strategies to provide natural lighting. Studies have shown that it has a positive impact on productivity and well being.
- Install high-efficiency lighting systems with advanced lighting controls.
- Use a properly sized and energy-efficient heat/cooling system in conjunction with a thermally efficient building shell.
- Maximise light colours for roofing and wall finish materials; and use minimal glass on east and west exposures.
- Minimise the electric loads from lighting, equipment and appliances.
- Consider alternative energy sources such as photovoltaic and fuel cells that are now available in new products and applications. Renewable energy sources provide a great symbol of emerging technologies for the future.

(iii) **Water efficiency:** ED considerations in improving water use efficiency through building design include the following:

- Design dual plumbing to use recycled water for toilet flushing or a gray water system that recovers rainwater or other non-potable water for site irrigation.
- Minimise wastewater by using ultra low-flush toilets, low-flow showerheads and other water conserving fixtures.
- Use recirculating systems for centralised hot water distribution.
- Install point-of-use hot water heating systems for more distant locations.
- Use a water budget approach that schedules irrigation using the California irrigation management information system data for landscaping.
- Meter the landscape separately from buildings. Use micro-irrigation (which excludes sprinklers and high-pressure sprayers) to supply water in non-turf areas.
- Use state-of-the-art irrigation controllers and self-closing nozzles on hoses.

(iv) **Materials efficiency:** ED considerations in improving material use efficiency through building design include the following:

- Select sustainable construction materials and products by evaluating several characteristics such as reused and recycled content, zero or low off gassing of harmful air emissions, zero or low toxicity, sustainably harvested materials, high recyclability, durability, longevity, and local production. These products promote resource conservation and efficiency. Using recycled-content products also helps develop markets for recycled materials.

- Use dimensional planning and other material efficiency strategies. These strategies reduce the amount of building materials needed and cut construction costs. For example, design rooms on 3-foot multiples to conform to standard-sized wallboard and plywood sheets.
- Reuse and recycle construction and demolition materials. For example, using inert demolition materials as a base course for a parking lot keeps materials out of landfills and costs less.
- Design with adequate space to facilitate recycling collection and to incorporate a solid waste management programme that prevents waste generation.

The Government of India has established a council for promotion of building material and technology (see Case 9.3).

LEARNING ACTIVITY 9.3

Name three building materials that are environmentally friendly.

Note:

a) Write your answer in the space given below.
b) Check your answer with the one given at the end of this book.

9.4 ED FOR DEVELOPMENTAL PLANNING

Throughout history, millions of people have migrated from rural areas to urban areas in search of places to work and livelihood. However, the experience has been that the employment generation in the secondary and tertiary sectors of the economy in the urban areas are not enough to effectively absorb the surplus labour from the rural areas. But, the migrations result in heavy stress on urban infrastructure thereby reducing overall quality of life. As urban areas enjoy better connectivity, both transport and communication, a vicious circle starts. The process of migration continues till the city expands in size and congestion. However, at a certain stage, the expansion becomes financially, environmentally and humanely unsustainable.

Historically, the concept of *carrying capacity* (CC) has been developed out of descriptions of growth and dynamics of natural populations, and as such has been used as a basis for range and forest management practices. Carrying capacity is defined as maximum population that can be supported in a given habitat without permanently impairing the productivity of ecosystem(s) upon which that population subsists. For human society, carrying capacity can be defined as the maximum rate of resource consumption and waste discharge that can be sustained indefinitely in a defined planning region without progressively impairing the bio-productivity and ecological integrity. Ecosystem carrying capacity provides the physical limits to economic development.

The goal of planning, therefore, is to provide a desired array of quality of life elements through physical and social design of the human environment. In accomplishing this goal, the planner traditionally works within the limits of what is technologically and economically feasible and what is socially, politically and legally acceptable.

Present planning process, however, is restricted to the allocation of monetary resources to various economic sectors to increase the production levels. Increase in production is believed to enhance the quality of life, which is defined narrowly as the availability of per capita income. Neither the perspective plan nor the five-year or annual plan delineates, in quantitative terms, the impact on the state of ecological resources as a result of envisaged planning process. Thus, the present planning process overlooks the basic requirement of availability of natural resources as also human and institutional resources that form building blocks in the developmental process.

The validation of the mechanics involved in developmental environmental planning, based on carrying capacity, is desirable, though challenging. Its success largely depends on the participation of planners as also the beneficiary community.

9.4.1 A Framework

The evolution of human society can be attributed to the continuous process of development in every sphere of human activity. The process of development requires growth in flow rates of materials and energies. This places an ever-increasing demand on the production and assimilation capabilities of resources in the natural environment. The capacity of natural and human environments to accommodate or absorb change without experiencing conditions of instability and attendant degradation is a significant concern in view of current trends of urban growth and development.

Recognising that the quantity, productivity and regenerative capacity of resources are limited, a strong motivation exists for a conceptualisation of environmental CC as a basis for regional planning and environmental management. Scarce resources must be managed in the context of many competing demands, and the natural and human environments must withstand perturbations caused by changes in our social and economic activities.

Researchers have used the idea of a limit or upper boundary on population density as a conceptual tool in the management of natural communities. Many of the difficult problems of environmental management arise, though, in the urban-regional context. Since urban systems interface with natural systems, the concept of carrying capacity in this setting needs enrichment in definition and interpretation. Rather than a single fixed number being rigorously established as a population carrying capacity, it suggests that planners should be concerned about the resource limits and environmental factors that may act as constraints or damping forces in the dynamic interaction of population growth, related socio-economic activities, the resource base, the environmental quality and quality of life.

Thus, the basic elements involved in assessing the carrying capacity hinge on the interrelation between resources and processes, which convert these resources to desired outputs and waste products and residuals, which must be assimilated back into the resource base with future effects on quantity and quality.

Carrying capacity may then be seen as the ability to produce desired outputs (i.e., goods and services) from a limited resource base (i.e., inputs or resources) while at the same time maintaining desired quality levels in this resource base. For an open system, the definition allows for import of resources, goods, services and the export of production and residuals.

The four dimensions that are relevant to the estimation of carrying capacity are:

(i) The stock of available resources to sustain rates of resource use in production.

(ii) The capacity of the environmental media to assimilate wastes and residuals from production and consumption.
(iii) The capacity of infrastructure resources (e.g., distribution and delivery systems) to handle the flow of goods and services and resources used in production.
(iv) The effect of both resource use and production outputs on quality of life.

Based on these dimensions, human carrying capacity may be considered as the level of human activity (including population dynamics and economic activity), which a region can sustain (including consideration of import and export of resources and waste residuals) at an acceptable quality or life levels in perpetuity. It also follows that carrying capacity must be assessed and properly identified with dimensioned domains. The domains of key concern are space and time.

Spatial Domain

Regional delineation for analysis of carrying capacity must be based on consideration of the functional elements of three general types of regional space: the ecological space, the economic space, and the policy space. The study area for analysis within these three regional dimensions should fulfil the following criteria:

- The study area should be relatively easy to identify and separate from other units.
- The study area should be capable of being manipulated separately with a minimal effect on other units.
- The boundaries of the study area should remain within, or resemble as close as possible, the less discrete boundaries of the ecosystem so that broad management practices are applicable throughout the study area.
- The movement of materials across boundaries can be measured.

Temporal Domain

In the human environment, the temporal, i.e., time-frame, perspective (time-frame) is an important aspect of carrying capacity study. Economic mechanisms stimulate technological advances that lead to increases in carrying capacity over time (e.g., development of internal combustion engines generating lower emission levels per RPM will allow greater traffic volumes).

An important aspect of carrying capacity assessment is the problem of measuring the outputs of production processes. Production outputs may be divided into two classes: goods and services and wastes and residuals. Measurements for the former, known as social indicators, may be used to gauge the relative change in human needs and satisfaction levels. Measurements for the latter, known as environmental indicators, allow managers and decision-makers to determine how production activity affects the quality and assimilative and future productive capacity of the receiving environment.

Several indicators may be integrated into a single index for more complex conditions or components (e.g., air quality index). Indices at an intermediate level of aggregation may, in turn, be integrated into still more general indices (e.g., quality of life). In these contexts, the term *index* represents a measurement of some environmental or social component for which there is more than one indicator. This provides a basis for evaluating the performance of development plans in terms of the changes in carrying capacity indices. In developing environmental management strategies for the urban region, planners and decision-makers must continually assess the social and environmental implications of various proposals. Recognising and establishing the limits of economic activity could provide decision-makers with a workable approach to assessing the environmental viability of developmental proposals.

To reiterate, planning for development within the limits of carrying capacity recognises that humankind is dependent on the productive capacity of ecosystems, and therefore, a minimal level of ecosystem integrity is essential for human survival. Planning for sustainable development calls for trade-offs between the desired production-consumption levels through the exploitation of *supportive capacity* within its regenerative capacity and environmental quality within the *assimilative capacity* of regional ecosystem. The utilisation of carrying capacity, thus, requires a series of adjustments to reconcile competing operations in the developmental process through participation of various stakeholders.

The supportive capacity of a region is the capacity of the ecosystems to provide resources for various anthropogenic activities, i.e., human settlements/industries. The resource base of a region could be categorised into ecological and economic resources, transformational resources, infrastructure and distributive resources and socio-cultural and amenity resources. Pragmatic utilisation of these resources warrants the establishment of functional relationships between the resources and their present level of usages. The carrying capacity of an ecosystem is greatly influenced by the availability of resources and the manner in which they are utilised. The availability of resources at any particular time is the result of the interactions amongst the physical occurrence of the resource, and the quantum of requirement, as also the technological and managerial means of their exploitation.

Assimilative capacity is the maximum amount of waste discharge that can be allowed in an environment without violating the designated use. In operational terms, assimilative capacity can be defined as the ratio of the observed level of pollutant and the environmental standard stipulated by the concerned authorities. Assimilative capacity is governed by dilution, dispersion, phase transformation, deposition and absorption phenomena. These phenomena are relatively local in nature. Assimilative capacity, therefore, needs to be worked out for a defined geographical area.

What the foregoing discussion suggests is that the major elements of work in carrying capacity study include the following aspects:

- Estimation of supportive capacity.
- Estimation of assimilative capacity.
- Assessment of present quality of life, developmental aspirations, developmental plans and policies.
- Development of alternate scenarios and strategies for improving quality of life within the limits imposed by the carrying capacity.

The overall scope of work for estimating carrying capacity, thus, involves:

- Delineation of boundaries of the study region, taking into account the location of anthropogenic activities likely to affect estuarine ecology, boundaries of *talukas* and watersheds.
- Assessment of stock and demand for land, water, minerals, transformational (e.g., power generation and manufacturing), infrastructure (e.g., water supply and sanitation, health and education, transport and communication and socio-cultural resources).
- Assessment of environmental quality vis-à-vis standards for air, noise, water (both surface and ground), land, biology (both aquatic and terrestrial) components based on secondary data (including time-series, if available) supported by primary data.
- Assessment of residual assimilative capacity.
- Assessment of quality of life through questionnaire survey.
- Assessment of present activity levels in various sectors of economy, viz., forestry, fishery, agriculture, manufacturing, trade and commerce, transport, waste management, etc.
- Identification and quantification of sources of pollution.
- Short-listing of environmental and developmental concerns.
- Review of existing developmental plans prepared by concerned authorities.
- Prediction of business-as-usual scenario in terms of impact of developmental plans on demand for supportive capacity and determination of environmental quality.
- Identification of limiting resources.
- Assessment of developmental potential of the region.

- Recommendation of short-term and long-term policy/planning/technology interventions for enhancing carrying capacity.

9.4.2 Indian Examples

All development projects utilise natural resources in one form or the other and also generate wastes. The quantum of wastes generated depends on the level of technology and management skills used. Obsolete technologies result in wastage of resources to the extent of 35-40%. It is, therefore, necessary to develop alternate strategies for sustainable development in the context of a developing country like India. CC-based regional planning takes into account not merely the availability of natural resources but also the environmental status of the ecosystem so that the welfare of the human beings as well as the health of the life support systems are not impaired.

To demonstrate the utility of the carrying capacity based planning process, studies have been conducted in India in a few selected areas considered to be representative of similar problematic areas in the country, some of which are listed below:

- Ecologically fragile Doon Valley.
- Dakshina Kannada district, Karnataka State.
- Environmentally critical national capital region.
- Industrially and environmentally sensitive Tapi Estuary.
- Industrialised and resource-rich Damodar River Basin.
- Coastal area with multi-sectoral development of Greater Kochi Region.

These studies have helped in identification of limiting resources in the face of developmental aspirations of the people of the region and suggest priority areas of action based on analysis of indicators of carrying capacity for business-as-usual and preferred developmental scenarios.

Let us now illustrate a typical case of carrying capacity based development study.

Tapi Estuary Study

In the last two decades, Gujarat, the tenth most populous state in India, has witnessed rapid industrial and economic growth. It is the second most urbanised state, with 34.5% population living in urban areas in 1991. The economy of Gujarat has traditionally been based on agriculture. In the recent past, this has been undergoing change with rapid industrialisation. Baruch and Surat are the major investment centres in Gujarat, and projects with total investments to the tune of INR 1,15,000 crores are under various stages of implementation. The Gujarat

State is likely to become the most industrialised state in the country and would emerge as a regional leader of industrialisation in South Asia.

During the last decade, the Tapi estuary region in Gujarat has become an ecologically sensitive area due to major industrial developments in the Kawas-Hazira region. The region is now the hub of India's petrochemical and fertiliser sectors with over large investments in major public, private and cooperative projects. The close proximity of Surat with Mumbai and good rail and road connections has resulted in rapid industrial growth of urban settlements, trading and other associated activities around Surat. Navsari, the adjoining urban centre, is also fast developing into another industrial centre. The rapid economic development has led to enhanced anthropogenic activities resulting in severe strain on urban infrastructure and natural resources of the region and hence the selection of the Tapi estuary region for the carrying capacity study.

Till 1996, the influence of tidal water and salinity used to extend up to Kakrapar weir, which is about 85 km upstream of the mouth of the Tapi river. It is, therefore, presumed that Tapi estuary water quality would be affected by all anthropogenic activities between Kakrapar weir to Tapi mouth. The eastern boundary of the study domain was set at Kakrapar weir. The north and south boundaries were set at banks of rivers Kim and Ambika river, respectively. This resulted in an area of about 6083 sq. km as study domain. From a functional point of view, the study domain can be divided into western coastal zone with Surat city as the urban node, central zone with predominant agricultural activity, and eastern zone, which is predominantly forest area.

During the formulation of the study, it was realised that the major resource problems in the study area relate to the changing land use whereas the environmental problems are due to improper wastewater management and air pollution. It was also understood that the concerned authorities have restricted the pace of planned industrialisation in the coastal zone of the study region. However, commercial activities and small-scale industrial units are increasing at alarming rates in the central zone, mainly along the transportation routes. A *prima facie* assessment also brought out disparities in terms of infrastructure and quality of life in the three zones of the study region.

In short, the central theme for operationalisation of the concept of carrying capacity for regional environmental management and basis for developmental planning was identified as:

- estimation of indicators of carrying capacity for economic activity, resource endowment, environmental quality, amenity, and quality of life;
- identification of aspirations of the people for development (improvement in quality of life);

- identification of limiting factors/critical resources;
- delineation of policy, planning, and technological interventions for augmenting the carrying capacity to meet the enhanced quality of life.

LEARNING ACTIVITY 9.4

Name five indicators you would like to use for planning industrial estate based on carrying capacity considerations.

Note:

a) Write your answer in the space given below.

b) Check your answer with the one given at the end of this book.

SUMMARY

In this Unit, we discussed the concept of environmental design (ED) or design for the environment. We began the Unit by giving an overview of the principles underlying ED and the benefits of adopting ED. We then discussed ED for manufactured products, construction and developmental planning. In this context, we introduced concepts such as eco-labelling, green building and carrying capacity.

REFERENCES

Belmane I,1999. *An Eco-innovation: cathode ray tube recycling at IBM Sweden*, Journal of Sustainable Product Design, Issue 9, April 1999, Centre for Sustainable Product Design, Surrey, UK

Brown, L.R. and Shaw, P. 1982. Six Steps to a Sustainable Society, World Watch Paper No. 48. World Watch Institute, Washington

Rowledge L.R., Barton, R.S., and Brady, K.S., 1999. *Mapping the Journey: Case Studies in Strategy and Action towards Sustainable Development*. Greenleaf Publishers, UK, 1999

SUGGESTED READINGS

Spiegel R and Meadows D. 1999. Green Building Materials: A Guide to Product Selection and Specification. John Wiley & Sons, Inc., New York.

Sustainable Building Technical Manual. *http://www.sustainable.doe.gov/freshstart/articles/ptipub.htm*

A Guide to Irrigation Water Needs of Landscape Plants in California. *http://www.dpla.water.ca.gov/urban/conservation/landscape/wucols/*

Department of Health Services, Indoor Air Quality. *http://www.cal-iaq.org/*

Khanna, P. and Kulkarni, V.S. 1991. An Approach to Carrying Capacity-based Developmental Planning Process. *Journal of IAEM,* Vol. 18, 12-19.

U.S. Department of Energy. *http://www.sustainable.doe.gov/buildings/gbintro.shtmll*

Environmental Building News. *http://www.buildinggreen.com/*

U.S. Green Building Council. *http://www.usgbc.org/*

CASE 9.1: CLEANER PRODUCTION PROGRAMMES IN INDIA

UNEP, 2002. Cleaner Production—Country Specific Compilations—Asia Pacific Region

Definition of Cleaner Production

In India UNEP's definition "Cleaner Production is a new and creative way of thinking about products and the processes which make them. It is achieved by continuous application of strategies to minimise the generation of wastes and emissions" is followed. Waste Minimisation is the primary driver concept.

Status on International CP Declarations and MEAs

- Not signed International Declaration on Cleaner Production (IDCP) as of December 2001; Signed and ratified the Vienna Convention (18/03/1991) and the Montreal Protocol (19/06/1992) on ozone depletion.
- Signed (10/06/92) and ratified (01/11/1993) the UNFCCC.
- Kyoto Protocol ratified in 2002.
- Signed (05/06/1992) and ratified (18/02/1994) the Convention on Biodiversity. Bio-Safety Protocol signed on 23/01/2001.
- Party to Basel Convention (24/06/1992).

CP Related Institutions in India

National Cleaner Production Centre was initiated in 1995 and hosted by the National Productivity Council, New Delhi. The centre has two core technical staff. Four Regional CP Centres:

- KCPC – under the Technical Consultancy Wing of the Department of Industries, Karnataka.
- GCPC – under the Gujarat Industrial Development Corporation, Gujarat.
- PCPC – under the Directorate of Industries, Punjab.
- WBCPC – under the Indian Chamber of Commerce in technical co-operation with West Bengal Pollution Control Board, West Bengal.
- Process cum Product Development Centres (PPDC) hosted by Ministry of Small Scale Industries, Government of India.
- Council for Scientific and Industrial Research (CSIR): National Environmental Engineering Research Institute (NEERI).
- Tata Energy Research Institute (TERI).
- Centre for Science and Environment (NGO).
- The World Bank as a part of pollution prevention programme has set up "Waste Minimization Circles" (WMC's) in different industrial units to demonstrate flow of enhanced economic returns in additions to compliance with environmental standards. The National Productivity Council, a semi-Government technical organisation and the host institution for the NCPC, has been appointed as the nodal agency. Around 60 WMC's have been established so far in the various regions of India focussing on specific industries, e.g., textile, metal finishing, chemical, tannery, and *Khandsari*.

Initiatives Taken to Promote CP

- Environmental Audit to report waste reduction policy statement on Abatement of Pollution.
- 'Eco-mark' Scheme to label environmentally friendly products.
- Establishment of Growth Centres and Integrated Infra-structure Development Centres.
- In 1993, a CP demonstration project DESIRE, targeting SME's was initiated by UNIDO, in cooperation with the Indian National Productivity Council and other industry associations. This DESIRE project focussed on three sectors: agro-based pulp and paper, textile dying and printing, and pesticides formulation.
- Urban and Environmental Infrastructure Facility in Karnataka and Rajasthan, promotion and commercialisation of renewable energy source (RES) technologies by strengthening the Indian Renewable Development Authority's capacity to promote and finance entrepreneurial investments in alternate energy.

- CTI is a USAID project implemented in partnership with ICICI Limited and U.S. Technical Assistance contractor—Tetra Tech. Clean Technology Initiative (CTI) that focusses on improving the environmental performance of companies in selected industrial sectors.
- The Centre for Science and Environment (CSE), an NGO, has initiated Green Rating Project (GRP) for industries. GRP is an attempt to present a market-oriented framework by which the environmental impact of industrialisation can be measured and monitored. As a reputation incentive programme it rates the environmental performance of companies within the specific sector. The sectors rated were the Pulp and Paper and Automobiles. CSE has also started Best Environmental Manager award, along with corporate policy assessment. *Source: http://oneworld.org/html/eyou/grp/grp_index.htm*
- A number of demo projects have been conducted either directly or through the Regional Cleaner Production Centres established by NCPC. 37 have been completed and eight are still going on.
- SIDBI in Karnataka has funded the development of a demonstration cell at a chemical plant in Mysore.
- UNIDO, in a joint initiative with the Government of India, Asian and Pacific Centre for Transfer of Technology (APCTT) and SIDBI, launched the UNIDO Technology Bureau for International Industrial Partnerships (TBIIP) in India. The Bureau advises participating institutions on project identification, appraisal and promotion, assists in formulating promotion strategies and identifies specific strengths and advantages of viable projects in order to upgrade their capacity in investment and technology acquisition from advanced countries and Indian technology transfer to selected developing countries.
- Under the APO's Green Productivity programme, National Productivity Council (NPC) of India demonstrated the application of GP concept in Dyes and Dye Intermediate and Tannery sectors.
- Life Cycle Analysis (LCA) has been adapted in the steel sector.
- Pollution Prevention/Cleaner Production related awards are initiated by industry associations like FICCI.

CASE 9.2: INDIAN INITIATIVES IN ENVIRONMENTAL DESIGN OF PRODUCTS: ENVIRONMENTAL INFORMATION CENTRE

Background

The Environmental Information Centre has been set up as a project under the Trade in Environmental Services and Technologies (TEST) programme of the United States Agency for International Development (USAID). EIC has been developed and is operated by FICCI's Business Information Services Network (BISNETINDIA) in collaboration with the Industrial Credit and Investment Corporation of India (ICICI) Limited, which is the implementing agency for the TEST project in India.

Objectives

- To generate awareness about environmental challenges.
- To stimulate proactive initiatives for improving environmental quality.
- To facilitate adoption of environment friendly technologies.
- To help create effective environmental management systems.
- To build synergistic partnerships towards sustainable development.

Activities

- Information compilation and dissemination through *www.Cleantechindia.com.*
- Studies on environmental issues, performance and technology options.
- Conferences and interactions for generating awareness, knowledge and stimulating new initiatives.
- Match making services for technologies and funding.
- Cluster-based technology assistance programmes for small industries.
- Audit and training service.

Online Information Services

- Indian Environmental Regulations: full online documentation.
- Guidelines and Manuals: compiled from reliable sources.
- Best Practice Guides: prepared by experts from around the world.
- Checklists and Tech-tips: for regular monitoring and trouble shooting.
- Success Stories: a large collection of case studies of key achievements.
- Technology Options: technology solutions for environmental problems.
- Sources of Equipment and Services: a global directory of over 8000 suppliers.
- Technology Sources: profiles of technologies and sources.
- News Briefs: daily news on environmental issues.
- Calendar of Events: schedules of forthcoming conferences, fairs, etc.
- Papers and Presentations: full versions in downloadable formats.
- Business Enquiries: offers and requests for doing environmental business.
- Global Web links: A classified list of other environmental websites.
- Directories: Environmental contacts in India and abroad.

CASE 9.3: INDIAN INITIATIVES IN ENVIRONMENTAL DESIGN OF BUILDINGS: BUILDING MATERIALS AND TECHNOLOGY PROMOTION COUNCIL OF INDIA

Background

In order to bridge the gap between research and development and large-scale application of new building material technologies, the erstwhile Ministry of Urban Development, Government of India, had established the Building Materials and Technology Promotion Council (BMTPC) in July 1990.

The Council strives to package proven innovative technologies for the benefit of entrepreneurs interested in setting up manufacturing units in tiny, small, medium and large-scale sectors.

The Council is structured to undertake the task of the extension and application of technologies and materials developed by research institutions on the ground with the backing of financial institutions and enabling regulatory environment.

The Minister of Urban Affairs and Employment is the ex-officio President and the Secretary, Department of Urban Employment and Poverty Alleviation, Ministry of Urban Affairs and Employment, the Vice President of the Council. The Board of Management of the Council have representations from the concerned Central Ministries viz. Finance, Industrial Development, Rural Affairs and Employment, Department of Power, Science and Technology, Small Scale Industries, Environment, Scientific and Industrial Research and Planning Commission. The State Governments, Housing Finance and Industrial Development and Promotion Agencies like HUDCO, NHB, IDBI, IFCI and ICCI are represented on the Board. To draw upon the experience and expertise in the area of building materials technology, eminent technologists, professionals of the public sector construction agencies, the corporate sector, R&D and standardisation organisations are also associated with the Board of Management.

Objectives

(i) To promote development, production, standardisation and large-scale application of cost-effective innovative building materials and construction technologies in housing and building sector.

(ii) To undertake such activities as required for promoting manufacturing of new waste-based building materials and components through technical support, facilitating fiscal concessions and encouraging entrepreneurs to set up production units in different urban and rural regions.

(iii) To provide support services to professionals, construction agencies and entrepreneurs in selection, evaluation, up scaling, design engineering, skill upgradation and marketing for technology transfer from lab to land in the area of building materials and construction.

Thrust Areas

- Improving the policy environment for sustained growth of low cost building materials, production and availability.
- Promotion of production units of building materials/components based on Flash, Red mud, Phosphogypsum, agricultural residues and other wastes and byproducts.
- Modernisation of small scale and village level building materials production units in rural and urban areas.
- Promoting economy in construction costs.
- Formulating of standards for local building materials.
- Strengthening industrial extension services for attracting more investment in building materials sector.
- Up scaling of technologies, know-how acquisition, absorption and dissemination.
- Global technology search and encouraging joint ventures in building materials and construction sector.

BMTPC's Activities

- Interact, assist and collaborate with R&D, standardisation and academic institutions and industries in India and abroad to ensure continuous development and promotion of alternate and cost efficient building materials and construction technologies.
- Identify and develop technologies and building materials based on agricultural and industrial wastes and promote proven technologies for rural and urban housing construction.
- Sponsor and assist financially, building material industries and related activities in urban and rural areas.
- Promote waste utilisation, wood substitution and plastics and polymers in building material and construction industry and encourage pilot plants and demonstration units/projects.
- Promote and assist building materials industries in production of construction components/systems and make recommendations to Government on matters of equity participation, venture capital support.
- Taking note of special needs of each region in the areas of building materials and construction technologies, undertake studies for forecasting technological requirement, social and environmental implications of various technologies and support further research and development.

- Persuade Central and State Government agencies, housing development and construction agencies and organisations in Private and Community Sectors for application of proven cost and energy efficient building materials and construction technologies.
- Advise entrepreneurs in technology selection, prototype development, commercial production and marketing and external appropriate support for development of processes/technologies and procuring equipment, etc.
- Identify, evaluate and undertake feasibility studies on innovative energy-efficient building materials/products and construction systems for extension of grass-root level through linkages with public, professionals, voluntary agencies and individuals.
- Review the various laws, fiscal provisions and pricing of building materials and advise Government in consultation with concerned Ministries/Departments on fiscal concessions, policy on subsidy, excise duty, tax concessions for promotion of appropriate building materials and technologies.
- Promote appropriate institutions in different parts to improve production and application of relevant building materials, skills and manpower development through Building Centres, Building Material Estates, Housing Guidance Centres and Artisans/Management Training Centres.
- Act as clearing house of technology transfer from lab to land, through production and application and act as repository of information on all types of building materials including conventional and newly developed, under development in India and abroad.

UNIT 10

Environmental Economics

STRUCTURE

OVERVIEW

Environmental economics, an emerging area of research in India, is assuming an increasingly important role in environmental management due to international initiatives/treaties. It is, therefore, necessary for us to understand the concept of environmental economics and related fields.

This Unit, accordingly, begins by introducing the relationship between economics and the environment. It then discusses the methodologies for assigning monetary value to tangible and intangible environmental costs and benefits. The Unit also discusses the economics of natural resources with particular reference to fisheries, forestry, water use and agriculture. It closes by introducing the concepts of regional economics and ecological economics. In short, the focus of the Unit is on the means of integrating economics and environment for making decisions related to developmental planning and environmental management.

LEARNING OBJECTIVES

After completing this Unit, you should be able to:

- use relevant economic terminology and explain the relevance of economics in environmental management;
- describe the techniques used to assess costs and benefits as they pertain to environmental factors;
- explain the use of market-based instruments for pollution control;
- discuss the concepts of regional economics and ecological economics.

10.1 ECONOMICS AND THE ENVIRONMENT

In this Section, we will discuss the relationship between economics and the environment. However, before we do so, we will first introduce you to the concepts of economics and financial assessment. An understanding of the difference between these concepts is essential because one of the prerequisites of carrying out a developmental project is the assessment of its economic feasibility and financial viability.

Economics describes the systems that humans have invented to facilitate the production, trade and distribution of goods and services to satisfy the needs of a community. It involves the exploitation of resources such as labour, materials, energy, water and land. The scope of economics is vast and goes beyond financial matters, i.e., dealings with money. Broadly, economic analysis can be categorised into macroeconomics and microeconomics.

Macroeconomics

This refers to large-scale operations of the economy of a nation, a group of nations or the world and deals with issues such as overall production of goods and services, unemployment, money supply and inflation, national debt and balance of payments. Statistics on these and other factors are used as indicators to gauge the progress of an economy and to compare economics of different nations. A central theme running through all macroeconomic analysis is the balance between public (government) and private economic activity and the degree of government control.

Microeconomics

Microeconomics deals with the operations of a particular industry or a sector of a national economy (e.g., mining, electronics). It is mostly concerned with prices and production processes. The literature on microeconomics, generally, contains supply and demand curves. Figure 10.1 below illustrates one such supply-demand curve:

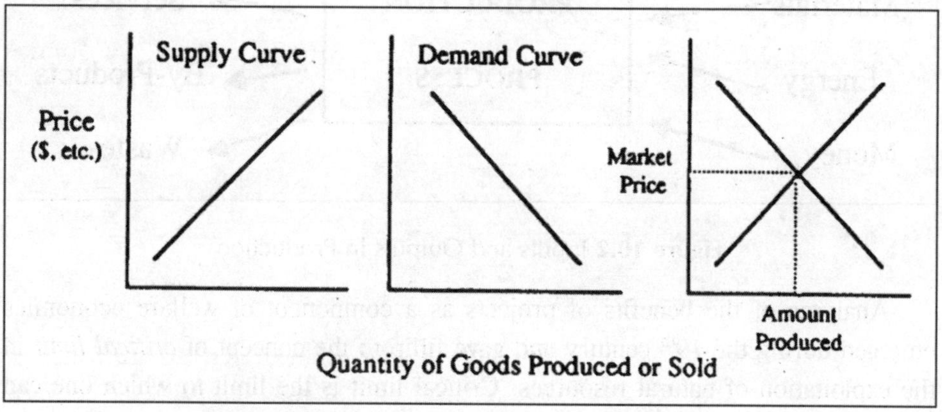

Figure 10.1 Supply and Demand Curves: An Illustration

As implied in Fig. 10.1 above, the basic elements of supply-demand curves are:

- The supply curve that expresses the attitude of producers of goods and services. (The higher the price for these, the more the producer will supply.)
- The demand curve that expresses the preferences of buyers or consumers. (The higher the price, the lower the quantity which will be bought.)
- Market price, which is an intersection point of supply and demand curves. (In a free economy, traders will be forced to bring the price and quantity produced to this equilibrium point, i.e., market price.)

Many microeconomic studies are concerned with optimising production by finding the most efficient level of output, the best combination of goods to produce or the best combination of inputs. The cost of inputs and the selling price of outputs are usually used as indicators of efficiency. The objective may be to maximise gross benefits or profits, minimise costs, or to maximise net benefits (i.e., benefits – costs). Microeconomic analyses of this nature are also relevant to projects such as the development of a factory or construction of a dam. Note that the level of production concerning large construction projects may be indicated by the production capacity, height of a dam or the number of lanes on a highway.

The production of goods and services represent an input-output process as shown in Fig. 10.2.

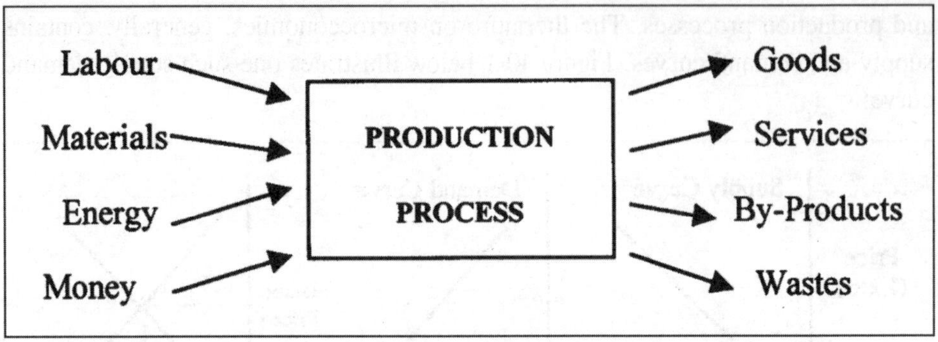

Figure 10.2 Inputs and Outputs in Production

Analyses of the benefits of projects as a component of welfare economics emerged during the 19th century and gave fillip to the concept of *critical limit* in the exploitation of natural resources. Critical limit is the limit to which one can exploit the resources without causing loss to other stakeholders. And, beyond this limit, any new development will disadvantage somebody. Economists, later, put forward a principle that a project was justified if its benefits exceeded the amount needed to compensate the persons who suffered losses, even when no compensation was actually paid. This theory was implemented practically in the methods of cost-benefit analysis.

A subset of economics is finance. But, it has its own procedures and philosophies. It is the domain of an accountant or a lawyer rather than that of an economist. It deals with diverse things by valuing them in terms of money. Put differently, funds are invested in developments and industrial activities, and income is derived from sales or fees for service. Records of expenditures and incomes must be kept in order that managers can control the operations efficiently and report to company shareholders and government regulators such as taxation offices, and provide a database for future planning. Facilities such as cost-control systems are, therefore, required and documents such as balance sheets and income statements must be prepared. There is also a legal responsibility for companies to prepare accounting documents in compliance with certain conventions.

Financial assessments can be performed in a similar manner to economic assessments, concentrating on financial cash flows. They usually involve the preparation of projected or proforma budgets, income statements and other documents.

> **LEARNING ACTIVITY 10.1**
>
> What is the difference between economics and finance?
>
> **Note:**
>
> a) Write your answer in the space given below.
>
> b) Check your answer with the one given at the end of this book.

Studies involving economic and financial matters are often carried out when assessing projects. These may be done either separately or as part of environmental studies as shown in Fig. 10.3.

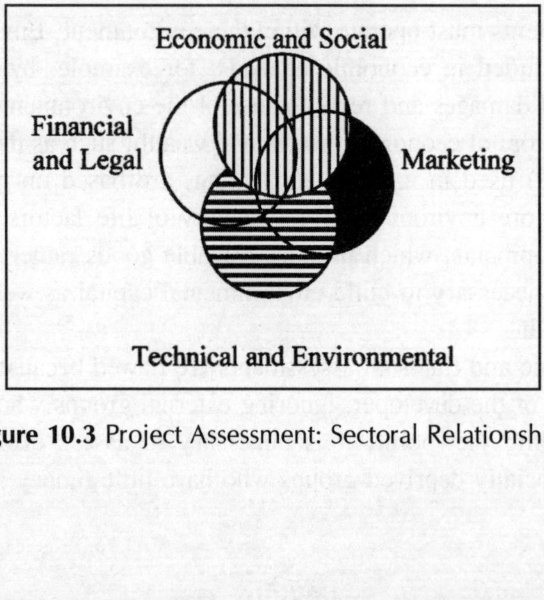

Figure 10.3 Project Assessment: Sectoral Relationship

How are environmental values to be considered in an economic context? We will discuss this issue in Subsections 10.1.1 to 10.1.4.

10.1.1 Environmental Economics: An Introduction

The study of economy-ecology interactions has gone far, even within the limits imposed by the discipline of economics. First, it is understood that economic activity uses natural resources and also returns waste to nature, and more importantly, the scale of that activity is determined independently of the rate of replacement of resources or waste absorption capacity of the environment. At low levels of economic activity, the asymmetry goes unnoticed because the inputs required for extraction are small. Traditional microeconomics treats environmental effects as externalities of production or consumption (Pigou, 1932; Ayres & Kneese, 1969). Beyond certain scales of economic activity, the pervasive nature of these externalities and limits on the waste assimilative capacity of environmental sinks results in the need for such residues to be treated as part of the materials balancing problem in the economy.

Environmental economics attempts to bring complex environmental factors into conventional economic analysis. At present, there are some problems with this. Since unintended environmental effects are secondary to the main purpose of a project, there is no market in which values can be set. Such effects are often called externalities, as owners or developers of projects see them as being outside their concerns. Also, there is disagreement about their importance or severity. Finally, as development has occurred at an unprecedented rate in recent times, and environmental concerns are relatively new, there is a lack of information on which to base judgements.

The new environmental economics emphasises that:

- Economic systems must operate within the environment. Environmental values should be included in economic analysis, for example, by defining costs of environmental damages and rents for use of the environment.
- Present indicators of economic growth and wealth, such as the Gross Domestic Product (GDP) used in national accounting, are based on market prices and commonly ignore environmental and quality of life factors. They are biased towards developments, which produce tradable goods rather than less tangible benefits. It is necessary to build environmental capital as well as human-made forms of wealth.
- Many economic and financial assessments are flawed because they reflect only the viewpoint of the developer, ignoring external groups who may be harmed. From the public viewpoint, it is necessary to assess effects on everyone, particularly socially deprived groups who have little money and influence.

- There is an issue of equity between the present generation and the future ones. Pursuit of short-term gains and consumption may cause lasting harm to posterity, especially if unique environments and resources are destroyed. It follows that renewable resources should be favoured over non-renewable ones. This question of intergenerational concern is one of the main arguments for sustainable development.

Optimal control methods analyse the constraints imposed on consumption growth by nature and the rates of regeneration of environmental resources. An extended analysis takes into account the multiple functions that environmental resources perform in relation to human survival and economic activity. The conditions of sustainable economic developments are derived in this framework. Further, if the resource stock has an amenity value and provides utility, the rate of extraction is more gradual and imposes limits on economic growth.

Barbier (1989) and Barbier & Markandaya (1989) suggest that an essential criterion for sustainable development is maximising the net benefits of economic development subject to maintaining the services and the quality of natural resources over time. A minimum viable level of the stock of environmental assets is built in as a constraint in addition to restrictions on rates of degradation, use of assets and levels of regeneration. This demonstrated that a low initial level of environmental quality results in environmentally unsustainable development. The discount rate makes the strategy of increased present consumption optimal. This implies that an economy with a high discount rate requires a higher initial level of environmental quality to avoid a growth path that is environmentally unsustainable.

Having introduced the concept of environmental economics, we will next discuss the techniques for valuation of environmental costs, benefits and also economic instruments (incentives and penalties) available for environmental management. But, before we do so, let us touch upon the concept of monetary values in Subsection 10.1.2.

10.1.2 Environmental Costs and Benefits

All assessments involve the identification of favourable and unfavourable effects. In economic terms, these are expressed as monetary costs or benefits and are estimated by using:

(i) past experience and results from activities similar to those proposed;
(ii) databases and processed information from bodies such as Statistical Bureau and Industry Associations;
(iii) informal information such as newspaper reports.

Costs

Costs of works can be determined from prices in past contracts and a company's cost accounting records. Market prices for goods can be found from records in statistical digests and trade journals. It is, however, likely that suitable data may not be available because a project is unprecedented, involves aspects that have never been tried before or is being implemented in a new environment. In such circumstances, effects and their costs and benefits must be estimated by special studies. These can be:

(i) Surveys of existing systems (e.g., a land survey of a planned highway route, an opinion survey of persons affected by aircraft noise from a new airport, etc.).

(ii) Models of non-existent systems (e.g., any system which is designed, but not yet built).

Costs can be classified in many different ways including the following:

- **Capital or first costs:** These are costs associated with the construction or implementation phase of a project, while operation, maintenance and replacement (OMR) costs relate to its ongoing operation. Sunk costs are costs incurred before the commencement of a project. Let us explain this further. For a project involving construction and operation of a large petrol station beside a motorway, the site preparation and construction of buildings will be among the capital costs. OMR costs will include the wages of employees, costs of petrol, electricity, etc. If the developer already owned the site prior to the decision to build the petrol station, it could be regarded as a *sunk cost*.
- **Direct costs:** These are costs attributable to some specific operation or component of a project, as opposed to general costs or overheads. For example, for cost-control purposes, a factory making 14 kinds of plastic products, could allocate costs to each product. However, items such as salaries of managers and office staff, water supply and building maintenance cannot be divided in this way, and must be regarded as overheads.
- **Variable costs:** These costs depend on the quantity of some items produced, unlike fixed costs that are independent of quantity (e.g., the cost of raw materials used by a shoe factory is variable, depending on its production, while the costs of advertising, rent and accounting are relatively fixed).
- **Marginal costs:** Marginal costs are the costs of adding one additional unit to a project. In the production of goods, costs vary with the amount of production, due to changes in the ratios of variable and fixed costs, shortages of inputs and efficiencies or inefficiencies at different levels of production. If total costs are plotted against project output or scale, the marginal cost is the slope of this

line. As production levels increase, marginal or unit costs decrease at first, due to economics of scale. They then reach a point of lowest cost, and then increase due to shortages of materials and other inputs.

- **Incremental costs:** These represent the differences in costs between two alternative projects or levels of project scale. Comparisons in cost-benefit analyses are made on the basis of incremental costs and benefits. For example, if a decision is made to connect two suburbs of a city by a light railway, and there are two possible routes, the incremental costs are simply the differences between the costs of these alternatives.
- **Opportunity costs:** These are the benefits foregone from some alternative investment when a particular project is adopted. For example, if irrigation water is to be supplied to a group of farms, the farmers will benefit from the increased production from conversion to irrigated agriculture. However, they will lose the ability to dry-farm the land. The real benefit is the difference between the benefits derived from irrigation and dry-land farming. Put differently, opportunity costs in this instance refer to the benefits from dry-land farming, which are lost.

We will next discuss a tool used in environmental economics, viz., environmental taxes. The use of taxation is an important tool for seeking environmental management goals.

10.1.3 Environmental Taxes

Capra (1997) suggested that one of the most effective ways of countering environmental damage and supporting sustainable development would be to shift the tax burden from income to *eco-taxes*. An eco-tax can be added to products, energy, services and materials to reflect true costs. This means that the consumer pays. While there have been national measures for some time, the interest in *green taxation* on an international scale is recent (and still mainly theoretical), which is triggered by increasing transboundary pollution, competition for internationally shared resources and the threat of global environmental change.

The function of green taxes is not to raise revenue for government but rather to provide participants in the marketplace with accurate information about true costs. For example, a tax on CFCs reflect their impact on ozone (Farber et al., 1995). Green taxes counter the pursuit of lower prices by externalising the true costs, ensuring that the purchaser is aware of the costs of environmental impacts. It is important that attempts to integrate external costs of production into prices do not burden the poor or punish the middle classes. The aim instead is to give people and companies incentives to invent, innovate and respond to environmental challenges (Repetto, Magrath., Wells, et al., 1992). Green taxation should encourage manufacturers to seek to reduce waste and other environmental damage to keep down their costs and thus prices to the purchaser, i.e., there is an incentive to improve environmental practice. Taxation is also becoming an important tool in the quest for sustainable development (Barrow, 1999).

There are a number of taxation approaches that have potential for controlling global climate change and these include trade able emission quotas; carbon (emissions) tax; energy use tax; taxation associated with technology transfers; reduced taxation for providing carbon sinks, etc. We will discuss two of them here.

Pigouvian Taxes

Individuals who consider only their own private costs and private benefits will do too much of any activity that generates negative externalities and too little of one that generates positive externalities (*www.tutor2u.net/economics/content/topics/externalities/policytowardsexternalities.htm*). When an activity generates both positive and negative externalities, private and social welfare will coincide only in the unlikely event that these opposing effects happen to offset one of these.

When externalities are present, the individual pursuit of self-interest rarely results in maximum social welfare. When it does not, we have an outcome that is, by definition, inefficient. This, in turn, means that it is possible to rearrange things in a way that makes at least some people better off without harming others in the process. There is an economic rationale for some form of government intervention in markets where externalities are prevalent.

How can we take into account the third party effects that necessarily arise? The key is to internalise the externalities that exist, i.e., make the firms and consumers that create the externalities take them into account when making their decisions. The classic way to adjust for externalities is to tax those who create negative externalities. This is sometimes known as making the polluter pay or introducing Pigouvian Taxes.

Carbon Emission Taxes

Ideally, a tax to control atmospheric emissions of carbon dioxide may be levied directly on the individuals or firms who are responsible for the emissions and may be based directly on the amount of carbon dioxide emitted. There are too many sources of emissions for direct measurement of emissions to be practicable. In practice, therefore, carbon taxes take the form of a tax on the carbon content of fuels, intended as a proxy for the carbon emission, which results from the combustion of these fuels. In some European countries (Sweden, Norway, Finland, The Netherlands and Denmark), where there is a policy of carbon tax, carbon taxes have taken the form of extended systems of fuel excises. Rates of taxes are defined separately for each fuel, in terms of fuel quantities, and relative tax levels on different fuels are set so as to equate the implicit rate of tax per unit of carbon across fuels. This requirement is, however, not always observed. In Denmark and Norway, for example, some fuels are not subject to the carbon tax. Also, the level of tax can vary across types of energy users. In Sweden and The Netherlands, for example, much lower rates of tax apply to industrial energy users than to energy use by private households. Most of the carbon taxes actually implemented in these countries have provisions to exempt firms or sectors, which are particularly exposed to international competition.

There are two schemes of carbon taxation: *primary carbon tax* (levied on primary fuels such as crude oil, coal and gas where they are mined, extracted or imported) and *final carbon tax* (levied on final fuel products such as coke, anthracite and four-star petrol sold to industrial users or households). Figure 10.4 illustrates the primary and final carbon taxes.

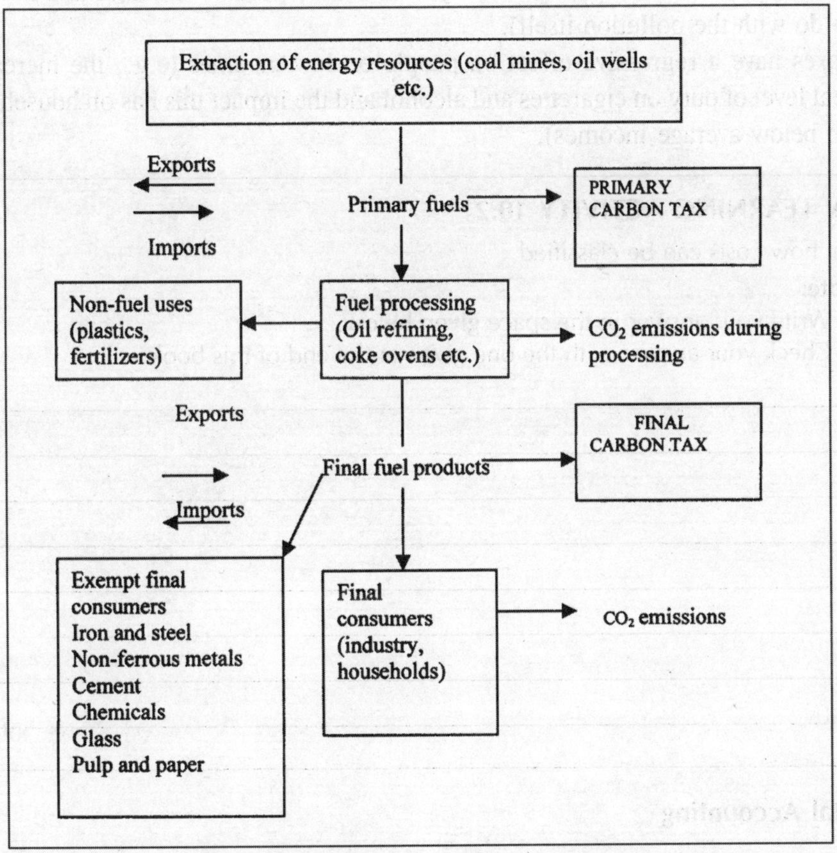

Figure 10.4 Primary and Final Carbon Taxes

Problems with Environmental Taxes

The problems with environmental taxes include the following:

- If taxation is too high, in part as a result of the problem of assigning accurate monetary values to the external costs created by producers and consumers, the outcome can be the expansion of gray markets where producers and consumers try to avoid the taxes. For example, one of the effects of the landfill tax has been an explosion in fly-tipping as producers seek to avoid paying the tax.
- Taxation may not be directed correctly, if it is hard to pinpoint precisely who is causing the pollution. This is the case along rivers where several industrial plants might be emitting effluents. Should producers be subject to a consistent tax regime when some are more at risk of polluting than others?
- Producers may be able to pass on the burden of the tax to the consumers, if the demand for the good is inelastic or the supply of the product is elastic.
- Higher taxes may cause cost-push inflation which itself may have detrimental effects on the economy (e.g., it may affect those people who have had nothing to do with the pollution itself).
- Taxes have a regressive effect on people on low incomes (e.g., the increased real level of duty on cigarettes and alcohol and the impact this has on households on below average incomes).

LEARNING ACTIVITY 10.2

List how costs can be classified.

Note:

a) Write your answer in the space given below.

b) Check your answer with the one given at the end of this book.

____

____

____

____

____

____

____

____

10.1.4 Environmental Accounting

Environmental auditing has been applied to eco-audits, stock taking, eco-review, eco-survey, etc. (Barrow, 1999). Environmental quality evaluations and environmental accounts systems collect data on the environment and resources to try and show the state of an area (or sea like the Baltic or Aegean). Most of these accounting procedures treat the environment as natural capital and try to measure its depletion or enhancement. Valuing the environmental features in monetary terms can be, admittedly, difficult.

The foundation for the accounting procedures has often been the UN model of Standard National Accounts, usually with *satellite accounts* added for environmental items. Some call these accounts as *environmentally adjusted national accounts*. These accounts seek to establish the stocks of resources, value of environmental features and their use over time. National environmental accounts systems, i.e., new systems of national accounts (green accounts, patrimonial accounts or state of the environment accounts) have been developed to assist with data gathering and storage and to value environment and natural resources.

Environmental accounting systems seek to set out a region's environmental, social and economic assets, and can be used to assess whether economic development is consistent with sustainable development, or to help ensure optimal use of natural resources and environment. For example, a natural resource accounting system can help the manager establish the percentage of, say, mineral exploitation profits to invest in long-term sustainability so that a region or country does not suffer a boom or decline. In practice, being able to make such investments depends on the type of government, people's attitudes and the persuasiveness of environmental management. Natural resource accounts can show the linkages between the environment and the economy, may be useful for forecasting and can establish the habitats, etc., that are of importance.

LEARNING ACTIVITY 10.3

List four areas where you will be required to use environmental economics as a tool for environmental management.

Note:

a) Write your answer in the space given below.

b) Check your answer with the one given at the end of this book.

____

____

____

____

____

____

____

So far, we discussed environmental economics and the various tools such as environmental taxes and environmental accounts to measure the value of the environment. We will next explain environmental valuation that helps us determine the economic importance of various impacts.

10.2 ENVIRONMENTAL VALUATION

The environmental problems the developing countries face are considerably different from those occurring in industrialised economies. For example, many rural populations in the developing economies depend on the direct exploitation of natural resources for agriculture, livestock raring, fishing, basic materials and fuel, both to meet their subsistence requirements and to sell in markets for cash income. Rapid land-use change has meant that many natural environments and habitats are disappearing quickly with the result that critical ecological resources and functions are being disrupted or lost.

The environment in developing countries is, therefore, very much related to health and welfare of rural and urban households. And, the basic production and consumption decisions of these households have a considerable impact on natural resources and environmental services. The lack of basic water supply, sanitation and other infrastructure services suggests that households highly value increased public provision of resource-based services. As a result, there has been a significant growth in the application of household production function (HPF) and production function (PF) approaches for valuing non-market environmental goods and services in developing countries. They have now recognised the importance of the environment to human welfare and economic livelihoods.

In Subsection 10.2.1, we will explain the need for environmental valuation in developing countries.

10.2.1 Categorising Environmental Values

In the developing countries, on an average, agriculture accounts for 40% of the GDP and nearly 80% of the labour force is engaged in agricultural or resource-based activities (World Bank, 1998). It is estimated that by 2020, the rural population of the developing world will have increased to around three billion and the urban population will double from 1.7 to 3.4 billion (Pistrup-Andersen et al., 1999).

Meeting the food needs of a growing and urbanising population with rising incomes will, in turn, put greater pressure on the agricultural resources of developing countries. Although food production is rising, the recent slowing in the growth in cereal yields has increased the demand for new land. Much of this land conversion will come at the expense of forests and other natural habitats. Annual tropical deforestation is 0.8%. Put differently, 15.4 million hectares are lost each year (FAO, 1993). However, land degradation is a severe problem in developing countries.

Around 20% to 25% of vegetated land in these countries suffers from some form of human-induced soil degradation (Oldeman, Hakkeling, and Sombroek, 1990).

Given these interlinking problems of population and environmental degradation coupled with the problem of poverty, there is a considerable interest among decision-makers in the developing countries in integrating environmental and economic policy concerns. As a result, there has also been an increased interest in the application of environmental valuation techniques. Valuation in developing countries has proven especially important in three areas: accounting for the cost of environmental degradation and damage; analysis of market and policy failures that contribute to environmental losses and public investment decisions with resource impacts (Barbier, 1994).

Economic values are classified either by type of value or measurement approach, and both are useful for categorising environmental values in developing countries. We will touch upon both these classifications, next.

Types of Economic Values

One pertinent feature of many tropical ecosystems and habitats (e.g., coastal and marine systems, rainforests, watersheds, range lands and floodplain wetlands) is that they exhibit multiple benefits or values. To inform the policies, these values are to be properly assessed. A framework for distinguishing and grouping these values is, therefore, necessary.

The concept of *total economic value* (TEV) is one such framework. TEV distinguishes between *use values* and *non-use values*, the latter referring to those current or future (potential) values associated with an environmental resource, which rely merely on its continued existence and are unrelated to use (Smith, 1987). Typically, use values involve some human interaction with the resource whereas non-use values do not. Table 10.1 gives an example of the TEV framework, as applied to a typical tropical coastal and marine system.

Table 10.1 TEV Framework: A Sample

Use values		*Non-use values*	
Direct use values	*Indirect use values*	*Option and quasi-option value*	*Existence value*
Fish Aquaculture Transport Wild resources Wood products Recreation Genetic material Scientific/educational	Nutrient retention/cycling Flood control Storm protection External ecosystem support Shoreline stabilisation	Potential future uses (as per direct and indirect uses) Future value of information	Biodiversity Culture, heritage Bequest values

As Table 10.1 suggests, use values are generally grouped according to whether they are direct or indirect. Direct use values refer to both consumptive and non-consumptive uses that involve some form of direct physical interaction with the resources and services of the ecosystem (e.g., harvesting of fish and wildlife resources, transport and use for recreation and tourism, etc.). Note that although tourism and recreation may be important in some areas, the most important uses of many coastal and marine systems in developing countries tend to involve both small-scale commercial and informal economic activity to support the livelihoods of local population, for example, through fishing, hunting, fuel wood extraction, and so forth. It is also increasingly being recognised that the livelihoods of populations in coastal and neighbouring areas may also be affected by certain key regulatory ecological functions. The values derived from these functions are considered to be activities that have direct measurable values (Barbier, 1994).

Economic Measurement Approaches

The measurement approach to the classification of economic values is usually determined by the following two criteria:

(i) If the valuation method is to be based on responses to hypothetical questions or on observed real world economic behaviour.
(ii) If the monetary values are to be inferred directly or through some indirect method of modelling individual behaviour and choice.

Table 10.2 illustrates the criteria for categorisation of economic values.

Table 10.2 Criteria for Categorisation of Resources for Valuation

	Observed behaviour	*Hypothetical*
Direct	*Direct observed*	*Direct hypothetical*
	Competitive market price	Bidding games
	Stimulated markets	Willingness to pay questions
Indirect	*Indirect observed*	*Indirect hypothetical*
	Travel cost/time allocation models	Contingent referendum
	Hedonic models	Contingent activity/behaviour
	Averting behaviour models	Contingent ranking/rating
	Agricultural household models	
	Random utility/discrete choice models	
	Benefit transfer approaches	
	Production function models	

Source: Freeman, 1993

An important step in environmental assessment is the valuation of the impacts (where feasible and appropriate) in monetary terms to determine their relative economic importance, and assess the benefits and costs of various alternatives. Let us discuss this in the next Subsection i.e. Subsection 10.2.2.

10.2.2 Valuation Techniques

Valuation techniques involve measuring the physical impact, and then assigning a value to that impact. A number of valuation techniques are potentially applicable to each category of value. The specific choice of technique will, however, depend on the situation and on data availability.

The valuation techniques comprise the following:

- **Valuation of changes in outputs and direct-costs and of changes in output of marketable goods:** In many cases, the environmental effects of projects manifest themselves (at least in part) in changes in output of marketable goods. Loss of forest, for example, results in the loss of timber products, fuel wood, fodder (whether collected or eaten on site by livestock grazed in the forest) and a variety of non-timber products such as fruit, herbs, and mushrooms. In cases such as these, the value of the unintended benefits and costs can be estimated by using the simple technique of valuing the change in output caused by the project. This approach is often referred to as the change-in-productivity approach.
- **Cost of illness and human capital:** Many environmental impacts such as air and water pollution have repercussions for human health. Valuing the cost of pollution-related morbidity (sickness) requires information on the underlying damage function (usually some form of a dose-response relationship) that relates the level of pollution (exposure) to the degree of health effect as well as information on how the project will affect the level of pollution.

  The cost of an increase in morbidity due to increased pollution levels can then be estimated using information on various costs associated with the increase in morbidity: any loss of earnings resulting from illness, medical costs such as for doctors, hospital visits or stays, medication and any other related out-of-pocket expenses. This approach is symmetric in that the benefits of actions that reduce the level of pollution and hence of morbidity are estimated in the same way. When this approach is extended to estimate the costs associated with pollution-related mortality, it is referred to as the human-capital approach.

  The human capital approach is similar to the change-in-productivity approach in that it is based on a damage function relating pollution to productivity, except that in this case the loss in productivity of human beings is measured. The human-capital approach is an extension of the more standard human capital theory, which relates the demand for education to its potential payoff in terms of expected lifetime earnings. Since, it reduces the value of life to the present value of an individual's future income stream, the human-capital approach is extremely controversial when applied to mortality.

- **Cost-based approaches:** When the benefits of a given environmental impact cannot be estimated directly, information on costs can be used to produce valuable information. For example, an order of magnitude estimate of the potential costs (or savings) to society from a change in an environmental problem can be obtained by using the cost of reducing or avoiding the impact, or the cost of replacing the services provided by the environmental resource.

  The major underlying assumptions of these approaches are that the nature and extent of physical damage expected is predictable (there is an accurate damage function available) and that the costs to replace or restore damaged assets can be estimated with a reasonable degree of accuracy. It is further assumed that these costs can be used as a valid proxy for the cost of environmental damage. That is, the replacement or restoration costs are assumed not to exceed the economic value of the asset. It simply may cost more to replace or restore an asset than it was worth in the first place. For example, cultivated hillsides may be eroding and there may be methods available (e.g., terracing, changes in cropping patterns) to reduce or prevent the erosion.

  However, each of these preventive measures has a cost, and it is the responsibility of the analyst to determine if the total costs of prevention are greater or less than the benefits of preventing erosion. In some cases, the costs of erosion control may be so high (and/or the benefits from controlling erosion may be so low) that erosion control measures will be an inappropriate use of scarce resources. In some other cases, there may also be more cost-effective ways to compensate for environmental damage than to replace the original asset or restore it to its original condition, and these substitution possibilities are ignored with the use of this technique.

  If substitutes are available, the method will be likely to overestimate the value of the damaged or destroyed asset. Due to this, these methods are generally thought to provide an upper-bound estimate of the benefits of measures taken to prevent the damage from occurring.

- **Replacement cost:** The replacement cost approach is often used as an estimate of the cost of pollution. This approach focusses on potential damage costs as measured by ex-ante engineering or accounting estimates of the costs of replacement or restoration if damage from pollution were to occur. For example, the costs of air pollution-related acid deposition in urban areas could be approximated by the restoration and replacement costs from damaged infrastructure.

  The replacement cost technique is particularly useful to assess the costs associated with damage to tangible assets, the repair and replacement costs, which are easily measurable. This information can then be used to decide if it is more efficient to allow the damage to occur and pay the replacement costs or to invest in preventing the pollution in the first place. The technique is less useful, however, for very unique assets such as historical or cultural sites and unique natural areas, which can neither be replaced nor easily be restored, and about which restoration costs are uncertain.

- **Relocation cost:** Similar to the replacement cost approach, the relocation cost approach uses estimated costs of a forced relocation of a natural or physical asset due to environmental damage.
- **Opportunity cost:** In some cases, it is decided to protect a particular resource and forego other development options. The term opportunity cost refers to the value of these lost economic opportunities due to environmental protection. It is, therefore, a measure of the cost of environmental protection in terms of development benefits foregone. The Global Environment Facility (GEF) and other donors may be willing to provide grant funds to cover these types of costs, especially when the benefits produced are important at the global level.

10.2.3 Valuing Environmental Amenities

Often, the environmental good or service being valued is not traded *per se* in the marketplace. Examples of these amenity-type services include recreational sites and the preservation of biodiversity. A number of valuation techniques exist that can be used to place monetary values on these resources and this information, in turn, can be incorporated into a more conventional benefit-cost analysis.

A number of tools can be used for valuing environmental amenities. We will touch upon some of them, next.

Hedonic Analysis

We know that environmental quality affects the price people are willing to pay for certain goods or services. Ocean/lake front hotels, for example, charge different rates depending on the view (e.g., rooms with ocean views cost more than the same size room with a garden view). Hedonic models have been widely used to examine the contribution of different attributes to prices for housing and to wage levels, including the contribution of environmental quality.

Many observed prices for goods are prices for bundles of attributes. For example, property values depend on the physical attributes of the dwelling (such as number and size of rooms, amenities such as plumbing, condition, etc.); on the convenience of access to employment, shopping and education and on a number of less tangible factors such as environmental quality. Since each house will differ slightly from others, the influence of the various factors on its price can be broken down statistically, provided sufficient observations are available. This approach is of interest because many environmental dimensions are likely to be embodied in property values. A home in a neighbourhood with low air pollution, for example, should sell for more than that in a neighbourhood with high ambient air pollution. Hedonic techniques allow this effect to be measured, holding other factors such as size and amenities constant.

In essence, the technique estimates the implicit prices for various attributes, which together make up the sale price. Hedonic methods require observations of the prices of goods and of the attributes of these goods. To enable the effect of the many different factors to be distinguished, large data sets are usually needed. Due to their data intensity and the need for open reporting of prices, the application of these techniques has had limited (but often successful) application in developing countries.

Travel Cost

Travel cost (TC) is an example of a technique that attempts to deduce value from observed behaviour. The TC technique was designed for and has been used extensively to value the benefits of recreation. It uses information on visitors' total expenditure to visit a site to derive their demand curve, i.e., the total benefits of the site's services. This technique assumes that changes in total travel costs are equivalent to changes in admission fees, and the demand curve is calculated, accordingly.

Note that the TC technique depends on numerous assumptions, many of which may not be tenable in the context of international tourism. For example, it generally assumes that travel cost is proportional to distance from the site and that people living at the same distance from the site have identical preferences. While these assumptions may be often valid in the case of national tourism, they may not be valid in the case of international tourism. This technique assumes a single-purpose trip and encounters difficulties, when trips have multiple purposes. It should also be borne in mind that the resulting estimates are site-specific. The main application of the TC technique in developing countries is valuing tourists' willingness to pay for national parks.

Contingent Valuation

Unlike the techniques that use observed data, the contingent valuation (CV) technique relies on direct questioning of consumers (actual or potential) to determine their willingness to pay (WTP) to obtain environmental goods. In principle, this technique can be used to value any environmental benefit. Moreover, since it is not limited to deducing preferences from available data, the technique can get information about the specific benefits of a proposed project. This also means that, with appropriately worded questions, the CV technique can provide an all-encompassing estimate of the perceived costs and benefits of environmental changes, in contrast to other techniques which, as noted above, often only provide a partial estimate of environmental costs and benefits.

LEARNING ACTIVITY 10.4

List the approaches to valuation of environmental impacts.

Note:

a) Write your answer in the space given below.
b) Check your answer with the one given at the end of this book.

10.3 ECONOMICS OF NATURAL RESOURCES

It is a rare day when the newspaper does not cover a topic relating to natural resources or the environment. The topics may be the supply of electricity and water to households, farms and businesses; shortages and high prices of petrol or diesel; debates about land use decisions; disputes between states over water rights; pollution from industries and vehicles, etc.

The bulk of the news about natural resources concern conflict over alternative ways to use them because the resources are scarce to serve everyone's wishes. Conflicts inevitably emerge when different people wish to use the same resource in different ways. For example, when a factory wishes to dispose pollutants into water used by villagers for drinking water, the central problem is that there is not enough to meet both parties' needs.

By definition, scarcity means that there is not enough to satisfy everyone's demand. This, in turn, means deciding not only who has rights over them but also how these decisions should be made. Economic analysis can identify productive patterns of natural resources use. It can assess the economic incentives that guide people towards or away from such patterns. It can also evaluate the implications of these patterns for equity and sustainability. This, in fact, is an essential tool for designing policy and management strategies.

Against this backdrop, in Subsections 10.3.1 to 10.3.4, we will discuss the economics of natural resources with particular reference to fisheries, forestry, water and agriculture.

10.3.1 Fisheries

Marine fisheries of the world are in trouble. Depletion of once-productive fish stocks, a sporadic occurrence earlier in the twentieth century, has progressively become a common event. The declaration of 200 miles as fishing zones by most of the world's coastal nations in the 1970s, which was a reaction to overfishing by foreign fleets, has had limited success in terms of conservation. To name one example, the Northern cod fishery in the Western Atlantic collapsed in 1992, in spite of intense management activity by the Canadian government. The collapse caused by extreme overfishing has destroyed the economy of the Province of Newfoundland, throwing some 30,000 fishermen and plant workers on to the unemployment roster. Only limited signs of recovery of this historical fishery had appeared by 1999.

Fisheries as Natural Capital

Modern economic theory treats renewable resource stocks as forms of natural capital. To describe a generic model, let $x(t)$ denote the biomass of a given fish population at time t. When the biomass is x, the net rate of population growth equals $F(x)$. Harvesting at the rate h reduces the net population growth rate accordingly.

$$\frac{dx}{dt} = F(x) - h \tag{10.1}$$

Here, the production of the natural capital, $F(x)$, is facilitated by the natural environment (driven ultimately by solar energy). Positive investment in natural capital occurs, if $h > F(x)$, whereas disinvestments occurs, if $h < F(x)$. Equilibrium at stock level is reached at $h = F(x)$. Figure 10.5 illustrates this phenomenon.

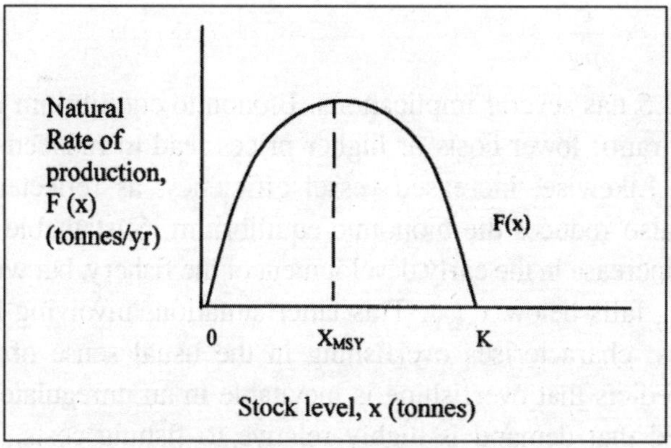

Figure 10.5 Natural Capital: An Illustration

In the fishery model, it is assumed that F peaks at some value $x = x_{MSY}$ and then decreases to zero:

$$F(0) = 0, F(K) = 0, F''(x) < 0$$

$$Max\ F(x) = F(x_{MSY}) = h_{MSY} \qquad (10.2)$$

Here K is the environmental carrying capacity of the population; $x = K$ is a stable equilibrium (when $h = 0$) of the natural population. The harvest rate h_{MSY} given by the Equation 10.2 is the maximum sustainable yield (*MSY*) that can be harvested from the population. *MSY* has been the traditional management objective for most fisheries, but achieving this objective in a stable, efficient and equitable way has proved to be more difficult than anticipated due to various reasons.

Next, we introduce the quantity E, the fishing effort (e.g., the number of standardised fishing vessels in operation at time t), and assume that:

$$h = qEx \qquad (10.3)$$

Here q, the catchability coefficient, may in general depend on the current stock size x. For simplicity, however, we assume $q = constant$. If c denotes the unit cost of effort, and p the landed price of fish, then the net flow of revenue (economic rent) equals:

$$R = pH - cE = (pqx - c)\,E \qquad (10.4)$$

On the basis of this simple model, we can discuss the concepts of bionomic equilibrium, depletion and optimal exploitation.

Bionomic Equilibrium

Bionomic equilibrium is the natural equilibrium point where the economic forces and the forces of biological productivity will be in balance. Consider an unregulated fishery exploited by numerous fishermen. If $pqx - c > 0$, that is, when:

$$x = x_{BE} = \frac{c}{pq} \tag{10.5}$$

Equation 10.5 has several implications. Bionomic equilibrium depends on the cost/price (c/p) ratio: lower costs or higher prices lead to reduced biomass levels at equilibrium. Likewise, increased vessel efficiency, as reflected by increased catchability q, also reduces the bionomic equilibrium. Sustainable yield equals F (x_{BE}). This will increase in the early development of the fishery, but will subsequently decline once x_{BE} falls below x_{MSY}. This latter situation, involving low fish stocks and low catches, characterises overfishing in the usual sense of the term. The model, thus, predicts that overfishing is inevitable in an unregulated, open access fishery, provided that demand is highly relative to fishing costs. The history of countless fisheries supports this prediction.

Externalities

Overfishing is sometimes attributed to externalities in the operation of individual firms/fishmonger competing in the fishery. In this context, externalities arise because each firm bases its decision solely on current returns to effort ($pqx–c$) and does not consider the effects of its catches on future stock levels. Additional short-term externalities may arise, if fishing vessels interfere with one another during search or capture activities. While such crowding externalities may influence the economic efficiency of fishing in important ways, they are probably less significant in terms of overfishing than the dynamic stock externality described above. Yet another form is the by-catch externalities, when firms seek a particular species (e.g., shrimp) or destroy other species (such as juvenile fish) that may also have value, either directly or as food for other targeted species.

Management Techniques

Traditional management approaches concentrate on controlling annual catches. Typically, the management agency determines a total annual catch (TAC). The cumulative year's catch is tracked and the fishery closed once that TAC had been reached. Alternatively, the length of the fishing season may be determined in advance, based on an estimate of the fleet's capture efficiency. Assuming that TAC has been correctly calculated and that actual catches are quickly and accurately reported, this method has the potential for protecting vulnerable fish populations. However, there are usually unfortunate economic side effects.

The restricted seasonal opening forces the firms to compete vigorously for their share of the TAC and also motivates them to increase vessel capacity so as to maximise their catches. Eventually, there will be pressure on the management agency to increase the TAC (or retain the current TAC) even though this may destabilise the fishery and precipitate a collapse.

Most fish populations undergo environmentally induced fluctuations from which they usually recover under natural conditions. When an overcapitalised fishery is dependent on the stock, such a fluctuation may result in a population collapse, unless the management agency responds quickly by reducing the TAC.

To prevent over capacity, management agencies may be forced to regulate vessel size, appropriate horsepower and fishing gear. They may also limit the number of vessels licensed to participate in the fishery. Such regulations tend to interfere with market-based individual decisions and lead to an increasingly unwieldy management system.

An alternative approach, now being used with increasing frequency, employs individual's transferable quotas (ITQs). Each firm has a specified annual quota, which may be caught whenever and however it desires. Quota units may be bought from or sold to other fishers. The management authority, as usual, determines total quotas. Provided that illegal catches can be prevented, the ITQ approach has many advantages arising from decentralised decision-making and the elimination of competition among firms. Economic rents are preserved and distributed among quota owners.

10.3.2 Forestry

The proper economic management of forests is one of the classic problems in renewable resource economics. To maximise land value, one has to choose the harvest time optimally. It is obvious that, if prices, the interest rate and the biotechnology remain unchanged, then each rotation will be of the same length over an infinite time horizon.

The Optimal Cutting Rule

Trees are biological resources in the same way as fish and crops, and the principles of forest management should, under well-defined property rights be analogous to what we have learnt in renewable resources (i.e., fishery). A fish stock should, in a steady state, be kept at a level where the yield from an extra unit added to the stock coincides with the real interest rate from an extra unit harvested and invested in a bank account.

This is a rule, which is analogous to the fundamental principle of profit maximisation in microeconomic theory: marginal revenue is equal to marginal cost. That is to say, the marginal cost (yearly) of keeping the fish in the fishing bank is the opportunity cost of not putting the market value of the fish in a bank. The marginal benefit (yearly) of an extra unit of the biomass is what it contributes to the stock over the period, and at the optimal steady state, it coincides with the interest rate. This is steady state criterion for the economically optimal stock. To preserve the stock at a steady state level, we have to harvest the yearly yield generated at the steady state level of the stock.

What distinguished trees from fish is that, for the latter population, it is neither practically feasible nor theoretically necessary to distinguish between fish in different age classes. Instead, we adapt the fishing tackle to catch the more mature individuals. However, it is both practically and theoretically crucial to keep track of the age distribution on forestry. In this case, it turns out to be more practical to focus on non-stationary conditions and decide the conditions for the optimal time to harvest an even-aged stand.

An approach, which can be used to find the optimal economic decision, is to consider marginal revenue and marginal costs. The marginal revenue from preserving the stand intact, until the next period, is the growth of the biomass over the period, or multiplied by the market price of timber. The opportunity cost of preserving the stand one more period is not only the interest on the revenue from selling the harvest, but also the interest on the value of the bare land. As indicated above, the latter component of the marginal cost of not harvesting the trees now surfaces, since one cannot start a new rotation without cutting down the standing trees. The marginal cost of keeping the stand intact can be decomposed into two components: the interest on the value of the standing trees and the interest on the value of the bare land.

We can sum up the discussion on the optimal rotation period under stationary external conditions such as constant prices, constant technology and a constant interest rate by the theorem proposed by Faustmann-Marshall-Pressler: a forest stand should be harvested when the change in its value with respect to time is equal to the interest on the value of the stand plus the interest on the value of the bare forest land.

Empirical Studies of the Timber Market

The empirical studies extend the discussion of forest management by formally introducing the concept of timber supply. To start with, if all prices, including harvesting costs, are increased by the same percentage, the rotation period remains unchanged. The intuition is that both marginal revenue and marginal cost increase by the same percentage, and they are, therefore, equalised at the original rotation period.

If only the prices for wood, but not the harvesting costs, are increased, the revenues from waiting yet another year, as well as the opportunity cost of the standing trees (interest on the value of the stand), will increase. However, the opportunity cost of the bare land will increase by even more. The reason is that revenues are scaled up, while costs are kept constant. To compensate for the increased marginal cost, the rotation period is shortened to increase marginal growth, and indirectly, marginal revenue. The immediate effect of a shorter rotation period is that less wood will, in the long run, be supplied from the stand. However, forestry has become more profitable, and it is likely that more land will be transformed into forestland. The long-term effect on wood supply is, therefore, ambiguous.

The Demand for Timber

The demand curve helps us to estimate both demand and supply curves from empirical data. The derivation of the demand curve can be dealt with very quickly. Timber and pulpwood are demanded as input in the forest industry. This means that we can use the theory of the firm and conclude that the input demand curve is downward sloping. We derive this result formally by a simple revealed-preference argument.

The Timber Market

Econometric analyses of the timber market require us to identify and estimate both the demand and supply curves, and identification is crucial. To identify the demand curve, we have to use the fact that the supply curve depends on entities such as felling costs, which do not affect the position of the demand curve. The principle is to use this shifter to cut out points along the fixed demand curve as illustrated in Fig. 10.6 where an increase in felling costs shifts the supply curve from S_1S_1 to S_2S_2 along the demand curve D_1D_1.

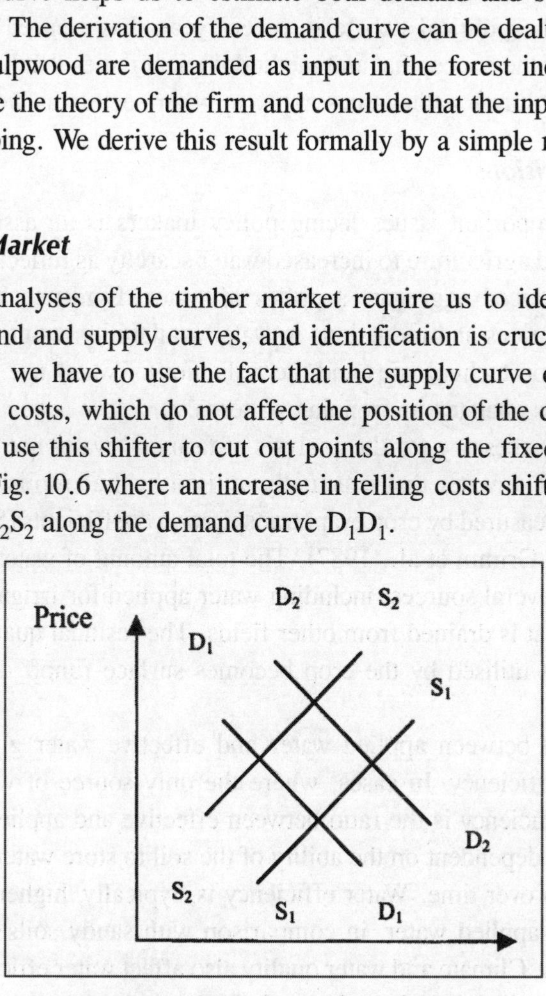

Figure 10.6 The Supply-Demand Curve

Along the rationale established in Fig. 10.6, we need a variable such as the price of lumber, which shifts the demand curve, to identify the latter.

10.3.3 Water Use

Water resources serve multiple purposes in meeting human needs and their allocation has been a subject of dispute over centuries. Rapid population growth has resulted in increased water scarcity and interest in the pursuit of improved productivity in water resource management.

On the global level, irrigated agriculture is by far the largest consumer of water among the various users, and concern over improving water use efficiency in this sector has been widely reflected in the water economics literature. The emphasis in this Subsection is on water use in agriculture and its impact on the environment because much of the political debate in recent years has centred on use in this sector, and many of the conceptual and methodological issues that have been raised in addressing agricultural water issues apply in other sectors as well.

Farm-level Water Allocation Decisions

One of the most important issues facing policy makers is an assessment of the response of irrigated agriculture to increased water scarcity as reflected in increased water prices or reduced aggregate supplies of water. Evidence has shown that increases in water prices and reductions in water availability result in the adoption of water conservation technologies and the reduction of water use within existing practices, as well as changes in cropping patterns.

A key concept necessary for the analysis of farmers' water use patterns is that of effective water. The water that is actually utilised by the crop is effective water and is commonly measured by crop evapotranspiration coefficients (Stewart, Hagen, Pruitt, et al., 1974; Grimm et al., 1987). The total amount of water applied to the crop comes from several sources, including water applied for irrigation, rainwater and runoff water that is drained from other fields. The residual quantity of applied water, which is not utilised by the crop becomes surface runoff or percolates to groundwater.

The distinction between applied water and effective water gives rise to the concept of water efficiency. In cases, where the only source of water is applied water, water use efficiency is the ratio between effective and applied water. Water efficiency is highly dependent on the ability of the soil to store water, which can be utilised by the crop over time. Water efficiency is, typically, higher on heavy clay soils, which retain applied water, in comparison with sandy soils through which water passes rapidly. Climate and water quality also affect water efficiency. Through their effect on water efficiency, these factors influence irrigation technology choices. To illustrate how land quality and effective water influence farmer's choice of irrigation technology and water use, we turn to a simple model.

Suppose a farmer produces a crop with the following per acre production function:

$$y = f(e) \tag{10.6}$$

where y denotes output per acre and e is the effective water. The production function f (*) has the regular properties of a neo-classical production function: $f(0) = 0, f'(*) > 0, f''(*) < 0$.

Let a_i denote applied water per acre under technology I, and α is the land quality index, assuming values from 0 to 1 from poor to good quality. For simplicity, assume two technologies: a traditional technology ($i = 0$) and a modern technology ($i = 1$). Modern irrigation technologies, which permit the application of small quantities of water over long periods of time, result in higher irrigation efficiencies, particularly on poor quality soils, which are incapable of retaining applied water. Thus, effective water under each irrigation type is a function of soil quality: $e = e_i(\alpha)$. Irrigation effectiveness is defined as:

$$h_i(\alpha) = e_i(\alpha)/a_i(\alpha) \tag{10.7}$$

and

$$1 > h_1 > h_0 > 0$$

The cost per acre associated with each technology is k_i. This cost includes annualised repayment of investment costs and annual operating costs. The modern technology is assumed to be more capital-intensive, so that $k_1 > k_0$.

The modern technology will be selected in cases where the increased profits from higher yields or lower water costs offset the higher costs associated with the adoption of the technology. These results indicate that modern technology adoption will increase with increasing water or output prices. In addition, modern technology adoption is more likely to occur with poor land quality, due to the high price of effective water under the traditional technology, and the land augmenting qualities of the modern technology.

The Economics of Groundwater Management

The optimal management of groundwater can be modelled as a non-renewable resource (e.g., desert regions with a closed aquifer) or as a renewable resource (i.e., when aquifers are recharged). The dynamics of optimal groundwater use depend on the price of energy, the interest rate, the efficiency of pumping and the hydrological characteristics of the system.

In the case of non-renewable groundwater with no technological change, the level of optimal pumping follows the standard hotelling rule. The rate of pumping declines over time with the interest rate, and the user cost of water (price minus extraction cost) rises. The introduction of technological changes, which increase water efficiency, leads to a reduction in extraction levels and extends the economic life of a finite aquifer. In contrast, improvements in pumping technology that reduce extraction costs actually tend to accelerate the pumping of aquifer.

In cases where groundwater aquifers can be recharged, the groundwater system may reach a steady state where pumping is equal to recharge. If the steady state level is below the initial stock of water, a decline in pumping over time will occur, until the steady state is reached. In cases where groundwater aquifers have already been significantly depleted, an initial period when replenishment exceeds pumping levels will precede the attainment of the steady state.

Groundwater resources usually exhibit the characteristics of common property goods, where externalities arise because current water users do not take into account the impact of their pumping on future water users. Current pumping affects future water users primarily by lowering the depth of water, increasing pumping costs and by reducing the stock of water available for future pumping. In the absence of any common-pool management scheme, current users determine their water use levels by setting the value of the marginal product of water equal to their private pumping costs, rather than full social costs. Correct situations of over pumping may entail establishing a regional or local groundwater authority either by taxing pumping, or by establishing aggregate pumping levels and introducing a system of transferable permits for pumping rights within the system (Gisser, 1983). Enactment of these situations requires monitoring of groundwater by a water agency, which may not always be feasible.

In many cases, the management of groundwater can be improved by removing policies which encourage overpumping, particularly the subsidisation of electricity. Electricity is the major cost component in operating irrigation pumps. Lower electricity rates decrease the cost of pumping and increase the depth to which it is profitable for farmers to pump. This is a major cause of over pumping and the depletion of groundwater aquifers. Frequently, ground and surface water systems can be managed conjunctively, where the optimal management of the system depends on the stocks and flows of the two water sources over time. Surface water canals have been established in many regions to compensate or augment depleted groundwater aquifers. In other situations, groundwater reservoirs have been established as buffer stock to supplement the use of surface water supplies in times of low supply.

10.3.4 Agriculture

Environmental externalities from agricultural activities, both past and present, are pervasive. Since Neolithic times, humans have been converting forests, wetlands and prairies into crop and grazing lands. These activities have shaped the rural landscape and the hydrology and ecology of agriculture as a bio-economic activity. Thus, agriculture is both a source and receptor of environmental externalities.

Contemporary evaluations of the environmental impacts of agriculture are both positive and negative. For example, the loss of biodiversity that occurred with the expansion of the agricultural frontier and human settlement in developing countries is viewed as an irreversible loss of natural capital. Yet, it is land drainage that has been an important factor in eliminating malaria in Europe and North America. In addition, agricultural land is an important habitat for many remaining wildlife species. Also, rural and urban populations often value agricultural landscape as open space.

Traditionally, agricultural policies attempt, with varying degrees of success, to achieve objectives related to farm income, agricultural prices and agricultural trade (Gardner, 1990; Johnson, 1991). Agricultural externalities, although influenced by the scale, location and methods of agricultural production, are at best a secondary consideration. Agricultural policy is concerned with encouraging the supply of positive agricultural externalities, and reducing the generation of negative externalities (Ervin and Graffy, 1996).

Negative Environmental Externalities

As is the case in any economic enterprise, energy and materials enter the farms as purchased inputs and environmental flows (e.g., rainfall, sun, etc.) and are removed as commodities and outflows to the environment. Moreover, while soils represent a stock of productive services, they can also be a flow of pollutants insofar as wind (air) and water carry eroded soils to off-farm locations, where they can cause significant environmental harm. Environmental outflows are in part the result of purposeful waste disposal activities such as burning fields to remove stubble or dumping unused pesticides in streams. However, natural processes such as runoff and volatilisation of nitrogen are important forces in the formation and fate of agricultural residues. Let us elaborate on this further.

Runoff from agricultural lands carry fertilisers, pesticides, livestock wastes, salts, pathogens and eroded soils into streams, lakes, estuaries and coastal waters. Many recent environmental assessments identify agricultural runoff as a major cause of surface-water quality problems in developed countries. Although the current knowledge of the state of groundwater is limited, there is evidence of groundwater contamination by agricultural chemicals around the world. Nitrates and pesticides that percolate into groundwater used to supply drinking water are of particular concern. The overall state of knowledge about chronic human health effects of pesticides is quite limited, but concern has been raised about the consequences of low exposures over long periods of time.

Economic costs from agricultural water pollution include reduced productivity of sports and commercial fisheries, in-stream recreational losses, increased cost of water treatment for industrial and domestic uses, costs of illness and death when contaminated waters are consumed, material damages and adverse impacts on the diversity and functioning of ecosystems. Although real or perceived human health rights often command the most attention, the costs associated with damages to the ecosystem and non-living systems are probably much larger.

Apart from reaching surface-water or groundwater supplies, excess nitrogen in the soil can be lost to the atmosphere through denitrification and volatilisation. Volatilised forms of nitrogen such as ammonia, nitrous oxide and nitrogen dioxide can contribute to acidification of the atmosphere and, in turn acid rain.

Agriculture, a source of several pollutants, is a contributor to greenhouse gas emissions. It is estimated that it accounts for one fifth of global emissions. The conversion of tropical forests to agricultural land, which releases carbon that had been sequestered in these forests, has perhaps garnered the most attention. Fossil fuels consumed in the process of using farm machinery and in the process of manufacturing fertilisers are other, albeit minor, sources of greenhouse gas emissions from agriculture.

Beneficial Environmental Externalities

Agriculture can be a source of a number of positive environmental externalities. One argument for agricultural land conservation policies is to safeguard agricultural production capacity from urban encroachment. The more cogent reason for agricultural land conservation policies, at least in countries with well functioning land and capital markets, is to influence the supply of related public goods (Bromley and Hodge, 1990). The study carried out by Kline and Wichelns (1994; 1996) shows that people are willing to pay to protect farmland from development. While markets can efficiently provide some of the uses of rural lands (e.g., crop production, hunting, etc.) and many others are inherently non-market goods (e.g., visual amenities, habitat preservation, etc.).

Hodge (1995) identifies the following three broad categories of land use policy objectives:

(i) To prevent undesired land use changes (e.g., urban containment).
(ii) To modify existing land uses to reduce detrimental externalities (e.g., encouraging farming practices that preserve wildlife habitat or discouraging practices that result in water pollution).
(iii) To stimulate new uses that give public or quasi-public goods (e.g., wetlands).

Agricultural Policy Change

A number of economic studies provide evidence that price supports, input subsidies and other agricultural policies influence the nature, size and spatial distribution of agricultural externalities through effects on the scale and location of production, input usage and structure. For example, policies that increase producer prices without restricting output (such as price floors, output subsidies, import restrictions) encourage farmers to increase production. Adverse environmental impacts occur when environmentally sensitive lands are converted to agricultural production and when there is increased intensity of the inputs that can increase environmental harm from agricultural production (e.g., pesticides, fertilisers, irrigation water and fossil fuels). Increased livestock production also means an increase in the volume of livestock wastes.

The presence of policies that distort agricultural product or input markets has implications for environmental policies for agriculture beyond their effects on the nature, size and location of agricultural externalities. For example, the costs of environmental policies that reduce surplus production or the excessive use of subsidised inputs will be overstated if the benefits from reducing these distortions are not counted. Shortle and Laughland (1994) demonstrate that the effectiveness of agricultural environmental policies can be diminished and the cost increased if distortion of agricultural policies are adjusted to compensate for the costs of compliance with environmental policies.

Policies for modifying agricultural land use to enhance rural amenities include regulations, cross compliance and environmental contracts. Architectural standards and other regulations can be, and are to some extent, used to control agricultural practices. However, in the provision of rural amenities, property rights and fairness considerations may reduce the feasible policy options to either voluntary action or subsidies to landowners for agricultural land use modifications (Hodge, 1989). Cross compliance makes eligibility for governmental benefits in one context contingent upon satisfying environmental performance goals in some other context.

LEARNING ACTIVITY 10.5

Suppose a proposal suggests that the existing solid waste dump located within residential area is to be shifted to a new location. List three costs and benefits that you would consider important in assessment of the proposal.

Note:

a) Write your answer in the space given below.

b) Check your answer with the one given at the end of this book.

____

____

____

____

10.4 ENVIRONMENT AND REGIONAL ECONOMICS

The emergence of regional economics as an established discipline dates back to the 1950's and aims to study the spatial patterns and processes of human activity from an economic perspective. Environmental economics has followed a slightly different developmental pace. Though the implications of external effects in a market economy were recognised, it was mainly the concern with the quality of life and that of the ecosystem that prompted interest in environmental economics.

Clearly, the concept of a region is central in regional economics. However, there is a multitude of regional demarcations: administrative regions, economic regions, natural resource regions and the like. A salient feature of a system of regions is its openness, which means the possibility of free trade, free movement of people and goods as well as transport of pollutants. The clear geographical focus in the region also means a close connection with the resource base and the pollution in a certain area.

We can, thus, conclude that there are many analytical connections between regional and environmental economic phenomena. These relationships may be unidirectional in nature, but may also show complicated feedback structures of modularly or hierarchically operating environmental economic systems in space and time. The nature of such interactions is dependent on ecosystems and human behaviour as well as on spatial environmental policy.

Spatial-environmental Policy Studies

Space can be observed in different ways:

- Space is the geographical medium, or physical market, for environmental externalities in a broad sense. This applies to global environmental change, but also to local issues like noise annoyance, soil pollution and so on.
- Space is of a heterogeneous nature with the consequences that environmental externalities have geographically discriminating and distributive impacts (e.g., recreational visits to attractive areas may pose an excessive burden on such areas).
- Space is a scarce good whose consumption (e.g., land use) has environmental implications and welfare effects for other members of society (e.g., environmental degradation processes in a global space-time context)—now or in the future, here or elsewhere.

In short, the space-environment relationship manifests itself as a complex nexus at the interface of regional economics and environmental economics.

Spatial-environmental policies cover a wide range of government measures such as local and regional taxation schemes or subsidies, land use zoning initiatives, regional environmental standards, industrial location strategies, infrastructure investments and the like. Such measures may be market-oriented or based on second-best principles. They may be of a control and command type or of a stimulating (discouraging) nature. Also, they may be strictly environmental in nature or address broader issues related to regional sustainable development. It is clear that different economic policies in an open economic system of regions have immediate implications for the environmental quality in each of the regions. Furthermore, different environmental policies in a spatial system of regions have consequences for the well being in each area. Thus, both regional-economic and environmental-economic policies may soon become competitive weapons in an open regional system. Spatial-environmental policy studies are subdivided into the following:

- **Impact studies:** These normally deal with 'what if' questions. Such studies are particularly important because in an open spatial-economic and environmental system, interregional flows (e.g., goods, services, migration, pollution, etc.) may lead to spatially discriminating distributive effect, which may affect the root of many conflicts. Clearly, both industrial policies and environmental policies may generate complicated spatial spill-overs, which may affect the welfare position of the regions at hand. For instance, public investment programmes, physical planning and infrastructure measures, land use zoning principles, establishment of regional environmental standards and the like will all affect the environmental quality of regions in an open spatial system.
- **Decision support studies:** These aim at mapping out the trade-offs between different policy options, including distributional conflicts and environmental quality consequences. In a spatial context, these trade-offs are also reflected in different development opportunities of regions or places in a spatial system. In several cases, however, research has to rely on ad-hoc information, expert opinion and so on, so that a precise assessment of the socio-economic and environmental implications of spatial-environmental policies is fraught with uncertainties.

LEARNING ACTIVITY 10.6

Can decisions for developmental projects be taken based solely on economic assessment of environmental impact? Justify your answer.

Note:

a) Write your answer in the space given below.

b) Check your answer with the one given at the end of this book.

We will next touch upon an emerging field in environmental economics, i.e., ecological economics.

10.5 ECOLOGICAL ECONOMICS

Ecological economics is a policy-oriented perspective that addresses the interdependence and co-evolution between human economies and their natural ecosystems. Interest in this area has been prompted by concerns of the adverse impacts of human economic growth processes on natural systems. Ecological economics recognises that humans and their economies are parts of larger natural ecosystems and co-evolve with those natural systems. There is a material and energy basis for the relations between human economies and their ecosystems, defining not only economic, but also social structures and processes.

The willful effort to extract useful things from natural systems is motivated by the satisfaction of basic biological needs and the seemingly limitless search for pleasure through consumption of goods and through social associations. The magnitude of potential impact on their own welfare through effects on natural systems requires that human decisions be guided by some notion of the value of their actions and the value of their impacts on ecosystems, either in terms of benefits of use or costs of abuse. Some concept of value is required for rational activities of human economies within their natural systems.

Both the structures and processes of natural systems have identifiable instrumental value to the human economy. However, natural systems also have aesthetic, moral and cultural values. These values are more intrinsic and immeasurable using traditional human preferences.

Valuation is made more complicated by the fact that our natural environment is more likely to shape values through establishing social and economic relations, aesthetic standards and culture. If so, our decisions now about the natural environment will shape future value systems, making values endogenous and, therefore, a poor guide to behaviour. A way out of this dilemma is to make valuations of natural systems based on what the society should be rather than on what the current valuations are.

Furthermore, our understanding of ecosystems is primitive. We know that structures change through normal succession and evolution, that processes are altered as the structures through which they change, that processes have various temporal and spatial scales and that catastrophic changes can occur without much evident alteration of structures and processes. However, beyond this abstract knowledge and except for a finite number of circumstances, we know too little about ecosystems to be confident that we can predict the full range of impacts of economy on ecosystems.

In order to understand the paradigm of an ecological economics framework, it is useful to contrast it to a characterisation of the current management paradigm. The two paradigms differ primarily on the primacy given to human economies versus natural ecosystems.

The prevailing management paradigm focusses on how humans can manage ecosystems for instrumental purposes of optimising human economic wealth. This wealth is, typically, measured in the value of utility enhancing things and actions, frequently measured by "willingness to pay" or "willingness to accept" monetary compensation for gains or losses and by summing across independent individuals. Preferences are, typically, taken as given and immutable and the manipulation of natural systems for human benefit addresses those preferences.

This management paradigm approaches uncertainty about natural systems by either denying or opting in favour of human economies. If not denying the uncertainty, the optimistic argument is given that natural processes are either reversible with enough time and engineering skill, or economic systems can find human-made replacements for lost ecosystem materials and services. The prevailing issue for this paradigm is that which relates to the use of ecosystem to effectively enhance human wealth and welfare.

The use of ecological economics requires knowledge of how ecosystems and economies work and how they interact at various spatial and temporal scales.

First, the ecological economics framework requires knowledge of both how natural ecosystems respond to economic activity as well as how economic activity responds to ecosystem changes. A seemingly useful analytical construct at this boundary is a full ecological-economic, input-output matrix. Flows of material, energy, nutrients, etc., between the economic and ecological systems will be quantified and impacts of one system on the other could be established.

Valuations of ecosystem services have, typically, been from the perspective of current generations and propose that value is represented by the willingness to pay for these services. The valuation methods developed by environmental economists may not be appropriate to valuing such services in a sustainability context. In a sustainability context, ecosystem structure and functions will be evaluated on the basis of the extent to which they contribute to the goal of economic and ecosystem health and sustainability rather than on the basis of their immediate contribution to current economic welfare.

Valuation of ecosystems based on individual preferences can be useful where spatial scales are narrow and temporal scales are short. However, the dramatic and potentially most serious ecosystem issues (e.g., global warming) are huge spatial and temporal scale problems. Preference-based valuations appear shallow in this context. For example, an appropriate question will be the willingness of future generations to pay current generations to avoid passing a legacy of a severely degraded ecosystem. The preferences of future generations are simply unknown, though possibly formable by current generations through education and cultural legacies and adaptable to future circumstances. The current generation's most empathetic valuation will consist of asking: How maximally sorry could the future be if we altered their inherited ecosystem by action X and what would the valuation of that sorrow be? It is difficult to find an answer to this question. But, we can pass on to the future an ecosystem that has the ecological properties of integrity and resilience, if an effective environmental/ecological policy is in place.

LEARNING ACTIVITY 10.7

In your opinion state how market-based instruments will be useful as an environmental management tool in the management of industrial estates?

Note:

a) Write your answer in the space given below.

b) Check your answer with the one given at the end of this book.

SUMMARY

In this Unit, we gave you an overview of environmental economics. We began the Unit by establishing the relationship between economics and the environment. We then presented some of the techniques for valuation of environmental costs and benefits. We also discussed the use of economic instruments for environmental improvements. In other words, we explained how the present standards could be met more cost-effectively. We also touched upon the concept of regional economics. Finally, we introduced the concepts of ecological economics to address the interdependence and co-evolution between human economies and their natural ecosystems.

REFERENCES

Ayres, R. U. and Kneese A. V., 1969.Productions, Consumption and Externalities, American Economic Review 59(3) 282-297.

Barbier, E., 1989. Economics, natural resource, Tech. rep., London: Earthscan Publications.

Barbier, E., 1994. The environmental effect of trade in the forestry sector. In: OECD (ed.), The Environmental Effect of Trade. OECD, Paris, PP. 55-101.

Barbier E. and Markendya A.,1989. Sustainable development and cost benefit analysis, London: Earthscan Publications.

Barrow C. J. 1999. Environmental Management: Principles and Practice, Rouledge, London

Bromley and Hodge, 1990 D. W. Bromley and I. Hodge, Private Property rights and presumptive policy entitlements: reconsidering the premises of rural policy, European Review of Agriculture Economics 17 (1990) PP 197-214

Capra, 1997. The Web of Life, Harper Collins Publishers, London

Ervin, D. E. and Graffy E. A. "Learner Environmental Policies for Agriculture". Choices (4th Quarter 1996): 27-33.

FAO, 1993. Forest Resources Assessment 1990: Tropical Countries, FAO Forest Paper 112, Food and Agriculture Organization of the United Nations, Rome.

Farber et al., 1995. "A Tax Incentive Tool for Environmental Management: An Environmental Scorecard" Ecological Economics vol.12 No 3 (1995): 183-189

Freeman, M.A.I., 1993. *The measurement of environmental and resource values: theory and methods*. Washington, DC: Resources For the Future.

Gardner, B, 1990. *The Economics of Agricultural Policies*, McGraw Hill: New York

Gardner, B L. 1990. The United States. In *Agricultural Protectionism in the Industrialized World,* ed. Fred H. Sanderson. Washington: Resources for the Future

Gisser, M. 1983.Ground Water: Focusing on the real issue. Journal of Political Economy, 91, 1001-27.

Hodge, I. D., 1989. Compensation for natural conservation. Environmental and planning A 27 (&) PP 1027-1036.

Hodge I.D., 1995. "Public Policies for land conservation" in the Hand Book of Environmental Economics, edited by D. W. Bromley Oxford: Blackwell.

Johnson, D.G.., 1991. World-Agriculture in Disarray. London: Macmillan & Co.

Kline, J. and Wichelns D., 1994. Using public referendum data to characterize support for purchasing development rights to farmland. Land Economics 70(2): 223-233.

Kline, J. and Wichelns, D., 1996. Empirical Evidence of public preferences for farmland preservation. In W.Lockeretz (ed), Environmental Enhancement through Agriculture. School of Nutrition Science and Policy, Tufts University, PP.265-274.

Oldeman, L. R., Hakkeling, R. T. A and Sombroek, W. G., 1990. In: Human Induced Soil Degradation International Soil Reference and Information Centre, Wageningen, the Netherlands

Pigou, A. C. 1932, The Economics of Welfare, 4th edition Macmillan, London

Pinstrup-Anderson P, Pandya-Lorch, Rosegrant M.W., 1999. The World Food Situation: Recent Developments, Emerging Issues and Long Term prospects, International Food Policy Research Institute, Washigton, DC.

Repetto, R., Magrath, W., Wells, M., Beer, C., Rossini, F. 1992, "Wasting assets: natural resources in the national accounts" in Markandia, A, Richardson (eds), Environmental Economics, Earthscan Publication, London

Shortle, J. S., and A. Laughland 1994. "Impacts of Taxes to Reduce Agrichemical use when farm policy is Endogenous" Journal of Agricultural Economics 45(1), 3-14 January 1994.

Smith, V. K, 1987. "Non-Use Values in Benefit - Cost Analysis" Southern Economic Journal 54, pp. 19-26.

Stewart J.I, Hagen R.M, Pruitt W.O, Hanks R.J, Riley J.P, Danielson R.E, Franklin W.T, Jakson EB, 1977. Optimizing crop production through control of water and salinity level in soil. Utah Water Research Laboratory, College of Engineering, Utah State Univ., Logan

SUGGESTED READINGS

Dixon, J. and Hufschmidt, M. (Editors). 1986. *Economic Evaluation Techniques for the Environment.* The Johns Hopkins, University Press, Baltimore.

Duncan Austin. 1999. Economic Instruments for Pollution Control and Prevention—A Brief Overview. World Resources Institute.

Munasinge, M. 1993. *Environmental Economics and Sustainable Development.* World Bank Environmental Paper 3, World Bank, New York.

Munashinge, M. and Schwab, A. 1993. *Environmental Economics and Natural Resource Management in Developing Countries.* World Bank, New York.

Oates, W.E. (Editor). 1992. *The Economics of the Environment.* Edward Elgar, Aldershot.

Pearce, D., Barbier, E. and Markandya, A. 1990. *Sustainable Development, Economics and Environment in the Third World.* Edward Elgar, Aldershot.

World Bank Environmental Department. 1991. *Environmental Assessment Sourcebook,* 2 volumes. World Bank Technical Papers No. 139 and 140. Washington, D.C.

World Bank. 1998. *EA Sourcebook Update No. 23,* Economic Analysis and Environmental Assessment.

Model Answers to Learning Activities

UNIT 1: PRINCIPLES OF ENVIRONMENTAL MANAGEMENT

Learning Activity 1.1

Some of the goals of environmental management are to prevent and resolve environmental problems; establish limits, set up and nurture institutions that effectively support environmental research, monitoring and management, identify threats and opportunities; sustain and, if possible improve, existing resources, improve the quality of life, etc.

Learning Activity 1.2

Population growth, resource depletion and degradation of environmental quality.

Learning Activity 1.3

Environmental management tools mentioned in this Unit are environment assessment (EA), economic assessment, usually through cost-benefit analysis (CBA), environment impact statements (EIS), environment audits, waste minimisation programmes and environmental management systems, life-cycle assessments (LCA) and environmental design.

Learning Activity 1.4

I am an Environmental Manager in an industry. The activities that may affect environment include transportation and storage of raw materials and products, water consumption in process and disposal of packing material.

Learning Activity 1.5

As a professional, the advantages will be:

(i) Co-ordinating activities involved in obtaining environmental clearance for expansion of my organisation.
(ii) Minimisation of waste generation from my organisation.
(iii) Development of environmental policy and programme for my organisation.

Learning Activity 1.6

Yes. It is possible to integrate environmental and developmental concerns. For example, small dams for irrigation, solar photovoltaic systems, use of biofertilisers, etc.

UNIT 2: POLICY AND LEGAL ASPECTS OF EM

Learning Activity 2.1

Product bans and extended producer responsibilities are two examples of policy instruments for cleaner production.

Learning Activity 2.2

The afforestation programmes in India include social forestry programmes, plantation along roads and rail lines, wasteland development, protection of Western Ghats and protection of Sunderbans.

Learning Activity 2.3

The five NGOs are:

(i) Council for Science and Environment.
(ii) Bombay Environmental Action Group.
(iii) Bombay Natural History Society.
(iv) Society for Clean Environment.
(v) Development Alternatives.

Learning Activity 2.4

My area of work is water so the treaties, environmental legislations and international conventions related to it are the Water (Prevention and Control of Pollution) Act of 1974, the Indian Penal Code of 1860, which penalises persons responsible for causing defilement of water of a public spring or reservoir with imprisonment or fines, the 1992 Helsinki Convention on the Protection and Use of Trans-boundary Watercourses and International Lakes, etc.

Learning Activity 2.5

(i) Ministry of Environment and Forest.
(ii) Central Pollution Control Board.
(iii) State Pollution Control Boards.

UNIT 3: ENVIRONMENTAL IMPACT ASSESSMENT (EIA)

Learning Activity 3.1

Environmental Impact Assessment (EIA) is a formal study process used to predict the environmental consequences of a proposed development project. Such projects may include, for example, building a hydroelectric dam or a factory, irrigating a large valley, or developing a harbour.

Learning Activity 3.2

- Selection of site for locating new industry which is not in an ecologically sensitive area.
- Selection of manufacturing technology which results in minimum environmental impacts.
- Involving project affected population in the process of project development.

Learning Activity 3.3

Country	*Law*
Thailand	Improvement and Conservation of National Environmental Quality Act 1975, amended in 1978.
India	Notifications dated May 5, 1994 under the Environment Protection Act, 1986.
Sri Lanka	National Environmental Act 1980, amended in 1986.

Learning Activity 3.4

Project	*Main Aspect Assessed*
Sardar Sarovar	Resettlement and rehabilitation of the project affected population.
Bombay Sewage Disposal Project	Impact of length on marine outfall on coastal water quality.
Chennai Water Supply	Impact of laying raw water pipeline on agricultural land project and forests.

Learning Activity 3.5

Highway project:

- Change in land use.
- Deforestation.

- Increase in air pollution.

Hydro electric project:

- Displacement of population in catchments area.
- Reservoir induced seismicity.
- Impact on terrestrial ecosystem.

Thermal power:

- Increase in ambient air pollution level.
- Increase in water demand due to cooling water requirements.
- Land pollution due to fly ash disposal.

Learning Activity 3.6

Stages	*Concerns*
Route selection	Deforestation.
Construction	Transportation and storage of road construction material.
Operations	Increase in air pollution and noise levels.

Learning Activity 3.7

Stakeholders in a highway project are:

- Land use planner.
- Department of road transport.
- Industrial development department.
- Persons affected due to land acquisition.
- Settlements along the proposed route or highway.

Learning Activity 3.8

The stakeholders that are required to be consulted, during public hearing of a coal-based thermal power project, are:

(i) Land owners whose land would be acquired.
(ii) Farmers in the surrounding area.
(iii) Fishermen who depend on fish yield of the waterbody from where the water would be extracted for the proposed unit.

Coal-based thermal power plant involves:

(i) Land acquisition for setting up the plant and ash disposal.

(ii) Impact on agricultural production due to the fly ash emission from the stack that increases the sulphur dioxide levels.

(iii) Impact on fish yields as a result of the cooling water disposal that increases the temperature of waterbody and toxic chemical levels (due to antifouling chemicals used in cooling water).

UNIT 4: EIA DOCUMENTATION AND PROCESSES

Learning Activity 4.1

The project proponent may not have the expertise for understanding environmental impacts associated with the proposed project. Also, he or she may be biased in favour of the project as proposed by him or her.

Learning Activity 4.2

- Size of the project (production capacity/investment).
- Location (ecological sensitivity).
- Pollution load addition.

Learning Activity 4.3

Power transmission line. This will help in the selection of route.

Learning Activity 4.4

No. The process of EIA has to take into consideration the socio-economic environment, topography of the proposed location and the existing environmental quality.

Learning Activity 4 .5

- Air: Winter
- Surface water: Summer
- Groundwater: Post-monsoon

Learning Activity 4 .6

- Optimal utilisation of time.
- Minimisation of cost of study.

Learning Activity 4 .7

- Identifying concerns of local people.
- Ensuring proper rehabilitation of project affected population.
- Ensuring co-operation of local people.

UNIT 5: ENVIRONMENTAL AUDITING

Learning Activity 5.1

Environmental statement helps the environmental manager to:

- work out resource consumption per unit of product;
- work out waste generation per unit of product;
- identify trends in resource consumption and waste generation;
- prepare plans for pollution reduction.

Learning Activity 5.2

The manufacturing process, product profile and workforce keep changing in the organisation. Also, after implementation of the recommendations of the audit, it is necessary to check the efficacy of these measures and devise a further action plan. An environmental audit should, therefore, be considered an ongoing process.

Learning Activity 5.3

Waste audit helps in the identification of the possibilities of waste reduction based on the study of the activities of the organisation.

Learning Activity 5.4

I am working in a textile industry. In my opinion a waste audit would benefit my organisation. In conducting a waste audit, I would include the following steps:

- Select process house limits for audit.
- Select audit team (we will have an internal audit team).
- Preliminary visit to the unit and discussions with the operators.
- Collect background information on raw materials used and products.
- Work out material balances.
- Compare raw material consumption and waste generation trend based on the records of last three years.
- Identify opportunities of waste reduction through good house-keeping.

- Identify equipment/process modification options.
- Prepare report prioritising the actions.

UNIT 6: LIFE CYCLE ASSESSMENT (LCA)

Learning Activity 6.1

The basic concept of life cycle assessment is that the entire life cycle of the product should be taken into account for evaluation of environmental impacts. An LCA of a product examines the product from the extraction of raw materials for the manufacturing process, through the production and use of the item to its final disposal.

Learning Activity 6.2

Vacuum or surface treated wood is the most environmental friendly product discussed in Annexure 6.2.

Learning Activity 6.3

Contribution to	*Unit*	*Coal*	*Heavy Fuel*	*Natural Gas*	*Wood*	*Straw*
Global warming	CO_2 emission, g/MJ	96	80	56	98a	99a
Acidification	SO_2 emission, g/MJ	650	500	1	33	110
Nutrient enrichment	NO_x emission, g/MJ	500	240	350	50	130
Smog formation	VOC, mg/MJ	15	7	15	1200	-
Chronic toxicity	Lead emission, mg/MJ	7.6	25	0.0	200	1.0

Coal is the fuel which is of more environmental concern.

Learning Activity 6.4

The four important steps in an LCA are:

(i) Definition of scope, goals and delimitation of the life cycle.
(ii) Preparation of an inventory.
(iii) Assessment of impact of environmental loadings in terms of environmental profiles.
(iv) Evaluation of environmental profiles according to the defined goals.

UNIT 7: ENVIRONMENTAL MANAGEMENT SYSTEM STANDARDS

Learning Activity 7.1

The use of environmental audit as a tool for environmental management assumes that an environmental manager is someone external to the industry. The aim of this audit is to facilitate an industry to comply with the regulations. Establishing an organisational system conforming to environmental management system (EMS) standards represent a paradigm shift away from the traditional approach based on command and control regulations, which set levels for pollutants. EMS focusses on management practices, which help companies assess their level of readiness and efficiency in meeting their environmental responsibilities.

Learning Activity 7.2

ISO follows three key principles in developing international standards, and these are:

(i) Consensus–the views of all interested parties are taken into account.
(ii) Industry-wide applicability–the goal is to draft standards that satisfy industries and customers worldwide.
(iii) Voluntary nature–all international standards developed are voluntary in nature.

Learning Activity 7.3

ISO standards are of two types: guidance and specification. Guidance standards are descriptive documents whereas specification standards are prescriptive documents. All the standards in ISO 14000 family, except ISO 14001, are guidance standards. Organisations do not become certified to ISO 14000 as a series, they certify to ISO 14001, the specification standard that is a model for environmental management system.

Learning Activity 7.4

As per ISO 14000, *environmental policy* is a statement by the organisation of its intentions and principles in relation to its overall environmental performance, which provides a framework for action and for setting of its environmental objectives and targets.

Environmental objective is an overall environmental goal, arising from the environmental policy, which an organisation sets itself to achieve. Where practicable, the objectives are quantified.

Environmental target, arising from the objectives, is a detailed performance requirement applicable to the organisation or parts thereof. In order to achieve the environmental objectives, targets are to be set and met. Where practicable, they are quantified.

Learning Activity 7.5

Implementing an ISO 14000 based EMS and using other tools from ISO 14000 families will give industries more satisfaction than merely complying with the prevailing legislations. The ISO 14000 approach forces an industry to take a hard look at all the areas where the industrial activities have environmental impact. This systematic approach can lead to the following benefits:

- Reduced cost of waste management.
- Savings in consumption of energy and materials.
- Lower distribution costs.
- Improved corporate image among the regulators, customers and the public.
- Framework for continuous improvement of environmental performance.

UNIT 8: ENVIRONMENTAL MANAGEMENT TECHNIQUES

Learning Activity 8.1

Input parameters: Noise level at source, distance of receptor from source.

System parameters: Temperature, humidity.

Output parameter: Noise level at receptor.

Learning Activity 8.2

We know:

$$r = \left(\frac{R_0}{1-(\frac{N_0}{K})}\right) = \frac{0.017}{1-(5.0\times10^9)/(15.0\times10^9)} = 0.0255$$

The time required to reach 7.5 billion, half of the final population size, can be found as:

$$t^* = \frac{1}{r}\ln\left(\frac{K}{N_0}-1\right)$$

$$= \frac{1}{0.0255}\ln\left(\frac{15\times10^9}{5\times10^9}-1\right) = 27 \text{ yrs}$$

To determine the number of years that it will take to reach 14.0 billion, we need to solve the equation for t:

$$N = \frac{K}{(1+e^{-r(t-t^*)})}$$

$$T = t^* - \frac{1}{r}\ln\left(\frac{K}{N}-1\right)$$

$$= 27 - \frac{1}{0.0255}\ln\left(\frac{15}{14}-1\right) = 130 \text{ yrs}$$

Learning Activity 8.3

The parameters for sensitivity analysis in prediction of impacts of combustion of coal in thermal power plant on surrounding air quality are:

- Ash content in coal.
- Sulphur content in coal.
- Stack height.
- Frequency of wind direction towards sensitive receptors.

Learning Activity 8.4

The probabilities that need to be considered in risk analysis of petroleum product storage tank in a refinery are:

- Rupture.
- Leakage.
- Explosion.

UNIT 9: ENVIRONMENTAL DESIGN

Learning Activity 9.1

The core objectives of environmental design are minimisation of use of resources including energy, during:

- manufacturing a product;
- constructing a project;
- use/operation of product/building;
- disposal of building/dismantling of building.

Learning Activity 9.2

The environmental considerations in the design of a washing machine are:

- extraction/production of material used for body;
- water requirement/kg of cloth;
- power consumption/hour;
- recyclability of the parts.

Learning Activity 9.3

The three building materials that are environmentally friendly are:

(i) fly-ash bricks;
(ii) particle boards;
(iii) aluminium frames.

Learning Activity 9.4

The five indicators that will be useful for the planning of industrial estates based on carrying capacity will be:

(i) air quality index;
(ii) depth of groundwater table;
(iii) water requirement per industrial plot;
(iv) power requirement per plot;
(v) percentage of working population in nearby localities.

UNIT 10: ENVIRONMENTAL ECONOMICS

Learning Activity 10.1

Economics describes the systems that humans have invented to facilitate the production, trade and distribution of goods and services to satisfy the needs of a community. *Finance* is a subset of economics, but one with its own procedures and philosophies. It is the domain of the accountant and the lawyer rather than the economist. It deals with diverse things by valuing them in terms of *money.*

Learning Activity 10.2

Costs can be classified as:

- Capital costs, operation and maintenance costs and sunk costs.

- Direct costs and indirect costs.
- Variable costs and fixed costs.
- Marginal costs.
- Incremental costs.
- Opportunity costs.

Learning Activity 10.3

Areas of application of environmental economics are the following:

- Construction of national highways passing through forests.
- Construction of a major dam.
- Selection of sites for industrial estates.
- Setting standards for pollution control.

Learning Activity 10.4

Various approaches to valuation of environmental impacts are:

- Total economic value.
- Direct use value.
- Indirect use value.
- Option value.
- Existence and bequest value.
- Benefit-cost vs. cost-effectiveness.

Learning Activity 10.5

Costs include:

- Cost of land acquisition.
- Monetary loss due to change in land use at new site.
- Monetary loss due to increase in transport cost.
- Monetary loss due to increase in pollution as a result of transport requirements.

Benefits include:

- Improvement in air quality in residential area.
- Improvement in water quality at present site.
- Monetary benefit through use of advanced technology such as power generation at new location.

Learning Activity 10.6

No decision about developmental projects can be taken solely based on economic assessment of environmental impacts. For complex proposals, it is unlikely that any economic analysis will provide a complete assessment. This is due to (a) uncertainty about the information on which analyses depend and the appropriateness of the methods and models they employ and (b) the boundaries imposed on their scope or extent. So, in most cases, we can view assessments - economic, environmental, technological and social - as contributing information to a decision-making process, rather than being the decision-making process itself.

Learning Activity 10.7

In industrial estates, there are large, medium and small-scale industrial units. While pollution control in large-scale industrial units may be technically and economically feasible, in small-scale units, it may not be possible. In such cases, market-based instruments will promote co-operation and collaboration among the units and thus, will be useful for reducing the total pollution load from industrial estates.

Distance Learning in Environment Engineering

Aimed at providing in-service professionals with the knowledge, skills and attitudes that enable them to recognise and solve problems in their domains of study and future work, Indian Institute of Science (IISc.) offers distance learning courses in Environmental Engineering.

The Institute uses various media for course delivery. The printed course materials, for example, are self-learning in nature, simulating learner-teacher interaction. Furthermore, appropriate information and communication technologies are used to facilitate just-in-time contacts among distance learners and teachers. The delivery model also incorporates an instructional system that uses performance-based learning processes and this enables you to demonstrate your understanding of the content-areas. The intensive competence-based course format allows you to take more responsibility for your own learning. In order to cater to the varying learning styles, the Institute offers the subject material in the following modes:

(i) Downloadable mode: Downloadable mode empowers to offer education in a dynamic form to students. Students can access lessons, assignments and submit their scripts online. They need not brood if they miss lessons - they can download their lesson notes and access archived lessons, tutorial sessions, lectures, etc.

(ii) Read-only mode: Students will be able to only read the course concepts, downloads or interactive sessions are unavailable.

(iii) Classroom interaction: The interactive learning provides an environment for students to talk and discuss freely about any related topics on or off line.

- Synchronous and asynchronous communication (X-chat, discussion forum, bulletin board, email, etc.): Client Server Model that works with Internet Relay Chat Server Protocol and allows servers to connect to each other effectively forming a network. Protocol is used by servers to talk to one another. Instructor and participants can talk with each other and discuss on related topics for a stipulated time given by the instructor. In mailing-group, participants can become a member of Mailing-group by registering their names in their group. Once approved by the instructor, participants can post their questions to this group ID. This way the course instructor or any other participant can respond to questions that will be read by all the members in that group including the instructor.

(iv) Practice orientation: Participants progress through the curriculum by demonstrating their competence in a variety of skill and knowledge areas. Student who demonstrates a competency is showing the ability "to do" something to put knowledge and principles into practice. After completing all examinations, students are required to apply the course content in his or her work setting and submit an example of how they achieved this goal.

Currently, the following two courses are on offer during January–May and August–December Sessions at the Indian Institute of Science, Centre for Continuing Education (*http://cce.iisc.ernet.in*):

1. **Environmental Management**
2. **Municipal Solid Waste Management**

Note that the course home page is the central location for course information. The faculty-in-charge of the course(s) will post their syllabi, grading policies, assignments, course schedules, etc., on the page. This apart, during the briefing session, you will

- learn about email functions,
- learn to use the Chat Room; an exciting facility that helps you communicate directly with your peers and the faculty at a distance,
- learn to communicate with your faculty in real-time,
- learn to view course grades in the Grade book,
- learn to use the Drop-box facility to submit and retrieve assignments,
- learn to make one-on-one communication between you and your faculty in connection with assignments,
- find out about the Document sharing, a public place in which you can upload documents to share and conduct "workshop",
- learn about the Online Help button for detailed, context sensitive instructions on how to use the tools.

Indian Institute of Science

(http://www.iisc.ernet.in)

The Indian Institute of Science (IISc) was started in 1909 through the pioneering vision of J.N. Tata. Since then, it has grown into a premier institution of research and advanced instruction, with more than 2000 active researchers working in almost all frontier areas of science and technology. IISc is an institute of higher learning and is constantly in pursuit of excellence. It is one of the oldest and the finest centres of its kind in India and has a very high international standing in the academic world as well.

The Institute currently has more than forty academic departments, pursuing research in different areas of science and engineering. The departments belong to one of the two faculties of the Institute, viz., the Faculty of Science and the Faculty of Engineering. The Faculty of Science awards PhD degrees through the Regular Research and Integrated PhD programme, while the Faculty of Engineering awards ME, MTech and MDes degrees in addition to PhD and MSc (Engg).

The Institute houses a huge collection of journals and textbooks in the J.R.D Tata Memorial Library. The Library also subscribes to many journals and publications in the electronic form. The Supercomputer Education and Research Centre makes available state-of-the-art computational facilities to the staff and students. The Centre runs round-the-clock, seven days a week and 365 days a year, helping the researchers to break many barriers in computational science and engineering.

The Institute has pioneered advanced education in India and has been making many significant contributions to frontier areas of research. The number of students in the Institute is kept deliberately small in order to focus on quality. It has been able to innovate and introduce (a) new systems of imparting knowledge and (b) educational reforms such as offering courses under unit system. The Institute was the first to introduce (i) Masters' programmes in engineering, (ii) the Integrated PhD Programmes in Biological, Chemical, Physical and Mathematical Sciences for science graduates, (iii) the new IISc Young Fellowship programme for the first 20 rank holders at the + 2 level and (iv) IISc Young Engineering Fellowship programme for merited III year BE/ BTech students.

The Institute has been primarily responsible for starting many fields of activity that include (i) Aerospace, Automation, Chemical, Civil, Computers, Electronics, Electrical, Mechanical, Metallurgical and Telecommunication Engineering in

Engineering and (ii) Biochemistry, Biophysics, Ecology, Materials Science and Solid State & Structural Chemistry in Science. It has also acted as a reservoir from which leadership and human resources for the future scientific developments and industries can be drawn. It is now involved in several emerging areas of national importance: space science and technology, environmental and atmospheric sciences, endocrinology, genetic engineering, rural technology and energy problems.

The Institute has a Centre for Scientific and Industrial Consultancy, through which significant R&D work is done on identified projects sponsored by industries. The know-how so generated has been transferred to industries. Transfer of technology has also taken place in areas such as low-cost housing and renewable sources of energy, which benefit the society. Similarly, the facilities available at the Institute such as low and high speed wind-tunnels, water-tunnel, high-speed computers and sophisticated instruments have been helping public and private sector industries, and defence. There has also been a certain amount of social utilisation of work in biosciences. For instance, (i) plant tissue culture of sandal wood, eucalyptus and teak wood, (ii) disease control in silk worms and (iii) nutritional value enhancement of rice strains. Some recent projects include the development of cryogenic equipment and vessels for advanced aircraft, fracture analysis of industrial and space launch vehicle components, thermo-metallurgical modelling of steel industries, acoustic absorbers to reduce acoustic pollution in industrial environment and automobiles and microbial techniques for gold extraction from mines. In 1997, the Institute released technologies relating to AIDS diagnostic kit, biomass gasifier and ferroelectric materials.

In all these endeavours, the Institute, with an awareness of its noble tradition and the need for maintaining a high quality in all its activities, strives to contribute to the scientific, academic and technological goals of India. Academic programmes leading to ME or MTech by course work and PhD and MSc (Engg.) by research are available in almost all the departments. A unique feature of the ME/MTech programmes at the Institute is the credit system which allows a student to tailor the Courses to suit his/her aptitude and interest. Every ME/MTech programme has a set of hard core Courses specified as an essential requirement whereas the student can take rest of the credits from many Courses available in his/her parent or other departments and also do a dissertation work on the topic of his/her choice. These Courses attract highly motivated, accomplished students, in addition to several sponsored candidates from R&D labs./industries and also from educational institutions under the Quality Improvement Programme (QIP).

A flexible approach to the ME/MTech programme has been recently introduced in a few departments, which enables candidates from industry to take Courses leading to ME/MTech, spread over a longer period than the regular programme, without having to leave their workplaces. Research programmes leading to MSc (Engg.)/PhD degrees are the main thrust in many departments. The programme has a limited amount of course work, essentially to prepare the student to carry out the research, but the main emphasis is on the thesis work. Excellent laboratory and computational facilities, a unique library, outstanding faculty with strong interests in both basic and applied sciences, make "dream come true" kind of environment to students with strong interest and aptitude for research. These programmes attract students from all corners of the country and beyond.

The External Registration program leading to MSc (Engg)/PhD provides a most attractive mechanism through which candidates from industry/R&D can work with the faculty at the Institute under the joint guidance of Institute faculty and senior officers, scientists in their parent organisation and acquire a research degree of the Institute with a minimal residential requirement. This has been one of the most successful programmes with nearly 30% research student enrolment in the Engineering faculty.

The Faculty being highly qualified and exposed in specialised areas of national importance attract a great deal of sponsored research and have the opportunity to participate at high-level national and international science and engineering meetings. This feature enables the faculty to respond rapidly to the changing scenario in science and technology and give initiations by introducing and nurturing new subjects. Thus, every research effort becomes indeed a step in a new direction, providing enriching experience of learning/discovering in the ever-changing scene of engineering and science. During the 1997-98 session, the Institute introduced three new Courses viz. (1) ME Programme in Microelectronics, (2) ME Programme in Signal Processing and (3) MDes Programme in Product Engineering and Design. To keep up with the changing scenario of science and technology development in the country, more Masters' Degree Courses are introduced since the 1999-2000 session, viz. (1) the 2-year MBA programme (which replaces the existing MTech programme) in the department of Management Studies and (2) the 2-year MTech (Computational Science) programme in the Supercomputer Education and Research Centre.

The Indian Institute of Science has for a long time been concerned with the ways to make its expertise and facilities available to qualified technical personnel in industries, universities and research establishments. On a small scale, the Institute's facilities are always being used through short-term/refresher courses, but these activities are not coordinated by a single entry. The Centre for Continuing Education (CCE) has therefore been set up in 1975-76 to function as a focus for all such activities and to promote a variety of programmes relevant to specific target groups (*http://cce.iisc.ernet.in*).

The Energy and Wetlands Research Group at the Centre for Ecological Sciences (CES), is actively involved in studies and training on issues related to environment, water resources, energy, ecology, wetlands, geographic information systems, environmental impact assessment and natural resource management. The information about research activities is available at *http://ces.iisc.ernet.in/energy/Welcome.html*

Recognizing the need for continuing professional development, the Commonwealth of Learning (COL) provided five high quality environmental engineering texts earlier developed through a UNESCO-COL partnership to the Energy and Wetlands Research Group at the Centre for Ecological Sciences (CES), Indian Institute of Science (IISc.) They jointly examined the relevance of, and the degree of modification required for, offering at a distance those texts for practising professionals in environmental engineering.

In collaboration with the Karnataka Environment Research Foundation, Indian Institute of Technology (IIT, Mumbai), National Environmental Engineering Institute (NEERI) with the financial support from the Commonwealth of Learning (COL), the Energy and Wetlands Research Group at the Centre for Ecological Sciences has developed five Environmental Engineering Courses in self-learning format suitable for continuous professional development. Based on the instructional problems, the audience analysis, instructional goals and objectives, and an understanding of the desired Course content, an outline of the content to be covered in the environmental engineering Courses were developed. Five Courses were developed simultaneously involving authors with respective domain expertise. The Courses developed in self-learning materials (SLM) for continuous professional development of engineers and the institutions are:

1. Municipal Solid Waste Management
2. Ground Water and Soil Pollution from Agricultural Activities
3. Air Pollution Control
4. Municipal Water and Waste Water Treatment
5. Environmental Management

These Courses are directed at practising professionals (in service professionals), engineers and technologists, in India, South Asia and the developing countries of the Commonwealth, who are all involved in environmental management in some form. The treatment, therefore, is not deeply technical or mathematical. In other words, building on their existing repertoire of skills base, through these Courses the target groups will be exposed to the new knowledge that is available in the field.

Next, existing materials along with outline of the content were reviewed by a programme review committee (at one week duration round table meeting) considering the relevance for distant learners with widely varied and non-traditional experiential backgrounds. The programme review committee chaired by COL included content experts, authors of various Courses, instructional design expert, project manager and a programme coordinator.

Instructional design was done based on content, need for instruction, audience need assessment, etc. Content was organised and developed using relevant examples from India that relate the content to a context understood by the students. This problem was addressed through discussions with a sampling of the target audience. The review of content and design was done at six months interval (mid term review and pre final review).

Evaluation was done including both formative and summation approaches to determine if the instructional methods and materials are accomplishing the established goals and objectives.

- Formative evaluation focussed on course strengths and weaknesses, technical or delivery concerns, and content areas in need of further coverage.
- Summative evaluation was done through validation and pilot testing. Pilot testing included both brainstorm ways and open discussion of Course contents. This provided a data for Course revision and future planning.

Within the context of formative and summative evaluation, data are collected through quantitative and qualitative methods combining quantitative measurement of student performance with open-ended interviewing and non-participant observation to collect and assess information about attitudes toward the Course's effectiveness and the delivery technology. Careful analysis of pilot testing results helped in identifying gaps or weaknesses as well as strengths and successes in the instructional process.

Revision plans typically were a direct result of the evaluation process in tandem with feedback from validation exercises involving Course authors from and content specialists. Also, it was evident from the pilot testing exercises that the instructor's own reflection on Course strengths and weaknesses were the best source of revision ideas. Most revisions were minor, such as breaking a large and unwieldy instructional unit into more manageable components, increasing assignment feedback, or improving student-to-student interaction. In one Course, major revisions were implemented.

The development of delivery models required integration of print, CD, and internet technology in concert with face-to-face communication. The challenge here was to integrate delivery components, based on identifiable learner needs, content requirements, and technical constraints.

Distance Education System and Study Skills

Studies across the globe, comparing distance education to traditional classroom instruction, indicate that teaching and learning at a distance are as effective as traditional instruction, when the learning materials, method and technologies used are appropriate to the instructional tasks and when student-to-student interaction and teacher-student feedback are robust. The typical challenges posed by distance education are countered by opportunities to:

- reach a wider student audience,
- meet the needs of students who are unable to attend on-campus classes,
- involve outside speakers who would otherwise be unavailable, and
- link students from different social, cultural, economic, and experiential backgrounds.

These days with advances in science and technology a wide range of technological options are available to the distance educator. They can be grouped into four major categories namely voice, video, print and web (Internet) options. Voice tools include passive or one-way technologies (tapes, radio, etc.) and the interactive technologies (telephone, audio conferencing, etc.). Video tools include still images such as slides, pre-produced moving images (e.g., film, videotape), and real-time moving images combined with audio-conferencing (one-way or two-way video with two-way audio). Print version in self-learning format is a foundational element of distance education programmes and the basis from which all other delivery systems have evolved. Various print formats that could be used for distance learning are: textbooks, study guides, workbooks, course syllabi, and case studies. With the advance and affordability of communication technology, web or Internet based learning become a reality. Technology plays a key role in the delivery of distance education. However, the focus of the programme is on the instructional outcomes rather than on the technology of delivery. In other words, the main thrust is on the needs of the learners, the requirements of the content and the constraints faced by the teacher, before selecting a delivery system. Typically, this systematic approach has resulted in a mix of media, each serving a specific purpose as touched upon below:

- Print component provides much of the basic learning content in the form of learning materials (which are in self-learning format), supplemented by suggested readings and other support materials in print.

- Interactive sessions provide real-time/face-to-face (or voice-to-voice) interaction (audio or video conferencing). This has proved to be, among others, a cost-effective way to incorporate guest speakers and content experts.
- Computer conferencing (e.g., chat session, discussion forum, bulletin board) or electronic mail to send messages, assignment feedback, and other targeted communication to one or more class members is also prevalent. This helps to increase interaction among students and between students and teachers.
- Pre-recorded video tapes/CD/DVDs are used to present class lectures and visually oriented content.
- Fascimile services are used, where affordable and required, to distribute assignments, just-in-time announcements, receive student assignments, provide timely feedback, etc.

Using this integrated approach, the task is to carefully select among the technological options depending on students (infrastructural facilities available in their locality), subject requirements (field visit, case studies, etc.) and institutional commitments. The goal is to build a hybrid instructional media, meeting the needs of the learner in a manner that is instructionally effective and economically prudent.

Distance education programmes have been initiated with careful planning and a focussed understanding of the Course requirements and student needs. Appropriate technologies have evolved through the efforts of many individuals and organisations, which rely on the consistent and integrated efforts of students, faculty, facilitators, support staff, and administrators. When instruction is delivered at a distance, additional challenges result because students are often separated from others sharing their backgrounds and interests, have few opportunities to interact with teachers outside the class, and rely on technical linkages to bridge the gap separating class participants. Hence, the success of distance education effort rests squarely on the shoulders of the faculty and resource persons. In a traditional classroom setting, the instructor's responsibility includes assembling Course content and developing an understanding of student needs.

NEED FOR ENVIRONMENTAL ENGINEERING DISTANCE LEARNING COURSES

Rapid industrialisation of developing countries has led to an increasing demand for continuing professional development for environmental engineering personnel at all sectors, including industrial and infrastructure management sectors. Recognizing the need for continuing professional development, the Commonwealth of Learning (COL) provided five high quality environmental engineering texts earlier developed through a UNESCO-COL partnership to the Indian Institute of Science (IISc.). They jointly examined the relevance of, and the degree of modification required for offering at a distance, those texts for practising professionals in environmental engineering. This resulted in a COL-funded project with IISc., taking the responsibility of coordinating and contributing to the design and development of a series of learning materials in Engineering Sciences for open distance learning.

Based on the analysis of target audience, learning goals and objectives, and the desired course content with particular relevance to India, an outline of the content was created, which was reviewed by a committee of content and instructional design experts. The development of five self-learning environmental engineering Courses included formative evaluation by the project review committee and electronic list discussions. The summative evaluation included validation and pilot testing and brainstorm sessions to ascertain the suitability of materials to meet the project goals and objectives. The materials thus produced were revised on the basis of the feedback and are on offer at a distance currently within India. Nonetheless, these Courses when suitably adopted with contextual examples can meet the continuous professional development requirements of the developing countries in the Commonwealth. It is also believed that these Courses can be appropriately modularised to contribute to the proposal put forward by the World Federation of Engineering Organisations for the World Engineering Partnership for Sustainable Development.

Course Content

COURSE 1 : MUNICIPAL SOLID WASTE MANAGEMENT

1. Municipal Solid Waste Management: An Introduction
2. Generation and Characteristics of Waste
3. Waste Collection, Storage and Transport
4. Waste Disposal
5. Waste Processing Techniques
6. Source Reduction, Product Recovery and Recycling
7. Recovery of Biological Conversion Products: Composts and Biogas
8. Incineration and Energy Recovery
9. Hazardous Waste: Management and Treatment
10. Integrated Waste Management (IWM)

Due to rapid increase in the production and consumption processes, societies generate as well as reject solid materials regularly from various sectors—agricultural, commercial, domestic, industrial and institutional. The considerable volume of wastes thus generated and rejected is called solid wastes. In other words, solid wastes are the wastes arising from human and animal activities that are normally solid and are discarded as useless or unwanted. This inevitably places an enormous strain on natural resources and seriously undermines efficient and sustainable development. One of the ways to salvage the situation is through efficient management of solid wastes, and this is the focus of this Course, Management of Municipal Solid Waste. In the ten Units that constitute this Course, we will discuss the processes involved in the management of solid wastes—from waste generation to final disposal.

In Unit 1, we will describe solid wastes and introduce you to the classification of solid wastes and the functional elements, such as waste generation, storage, collection, transport, processing, recovery and disposal, in the management of solid wastes. In Units 2 to 7, we will explain with the support of case studies each of these functional elements. In Unit 8, we will explain the treatment of solid wastes by incineration and energy recovery from the incineration process. Subsequently, in Unit 9, we will deal with the treatment and management of hazardous (biomedical) wastes. Finally, in Unit 10, we will discuss the concept of integrated waste management.

COURSE 2: ENVIRONMENTAL MANAGEMENT

1. Principles of Environmental Management
2. Principles of Ecology, Environment & Environmental Management
3. Policies and Legal Aspect of Environmental Management
4. Overview of Environmental Impact Assessment (EIA)
5. Environmental Audit
6. Life Cycle Assessment as EM Tool
7. Environmental Management Systems Standards: ISO 14000 (EMS)
8. Related Issues in Environmental Management
9. Environmental Design
10. Environmental Economics

Due to unplanned developmental activities as well as ever-increasing population, which has caused enormous strain on the environmental resources, societies across the world face several problems of environmental degradation. However, it is imperative to maintain a balance between the capacity of the environment and the quantum of sustainable utilisation. This is only possible by understanding the environment in its totality and the principles of its scientific management.

Environmental management (EM) has become one of the most used terms in recent times. But, what exactly does the term mean and entail? Different individuals belonging to different disciplines approach EM differently and therefore it is difficult to find a single, comprehensive view of EM. However, we will discuss in detail the various issues relating to EM in the ten Units that constitute this Course.

We will first discuss the fundamentals of EM and ecosystem in Unit 1 and, subsequently, the various environmental policies, legislations and international treaties in Unit 2. In Units 3 and 4, we will deal with the concept of environmental impact assessment (EIA) and the preparation of EIA report, respectively. We will discuss the methodology and processes of environmental auditing in Unit 5. In Unit 6, we will introduce you to life cycle assessment (LCA) in the context of EM. In Unit 7, we will explain the various EM system standards. We will then take up the issues and techniques relating to EM in Unit 8. Finally, we discuss environmental design and economics in Units 9 and 10, respectively.

COURSE 3: AIR POLLUTION CONTROL

1. Introduction to Air Pollution
2. Air Pollution: Impact, Legislation and Standards
3. Air Pollution Measurement

4. Prevention of Air Pollution
5. Air Pollution Meteorology
6. Air Pollution Modelling
7. Air Pollution Control: Particulate Matter I
8. Air Pollution Control: Particulate Matter II
9. Air Pollution Control: Volatile Organic Compound (VOCs)
10. Air Pollution Control: Sulphur Dioxide and Oxides of Nitrogen

A healthy person inhales about 16 kg of air per day. If the air is polluted, it will have serious implications on the health. Inhalation of polluted air considerably reduces our intake of oxygen, adversely affecting our health. However, on the one hand, we treat the atmosphere as a vast sink for air pollutants, and, on the other, we take clean air for granted and assume that it is a free and unlimited resource.

According to a World Bank study (2000), the number of premature deaths due to air pollution in India has increased by almost 30%. One of the reasons for this increase, the study reveals, is indoor air pollution. Air pollutants released indoors are likely to be more harmful to health because people are close to the place of combustion. In fact, indoor air pollution, caused by the combustion of non-commercial sources of fuels such as bio-fuels in homes that are poorly ventilated, kills thousands of young children and women annually. Put differently, children, women and the disabled staying indoor are subjected to pollution levels that may go up to 20 times the acceptable limits. Note that while almost three-fourth of Indian households use fuels such as firewood, crop leftovers, cow dung and other organic wastes, the remainder uses coal, mineral oil and compressed gases.

Note, also that as air pollutants get dispersed throughout the atmosphere, air pollution is not merely a local problem at the villages, large urban centres or industrialised regions. It, indeed, is a global problem.

Now, are there ways by which we control/prevent air pollution and ensure the present and future generations have abundant supply of clear air?

In this Course, Air Pollution, we will discuss the important issues pertaining to air pollution and its control/prevention. In the ten Units that constitute this Course, we will deal with the various sources of air pollution and the environmental and health impacts as well as the appropriate measures to control/prevent air pollution. We will begin the Course by discussing the sources and classification of air pollutants in Unit 1. Subsequently, in Unit 2, we will study the impacts of air pollution at various levels – personal, local, regional, continental and global – and environmental legislations and air quality standards. While Unit 3 considers air pollution measurement, Unit 4 discusses air pollution prevention through process, raw material and operational changes that can be brought about in process industries and other activities. We will then explain the meteorological aspects of air pollution in Unit 5 and list the different air pollution models and their application in Unit 6. In Units 7 and 8, we will discuss the different devices for the separation of particles from polluted gases. In Unit 9, we will consider air pollution control for volatile organic compounds (VOCs) in gases. In Unit 10, we will discuss the different processes and operations for the removal of sulphur dioxide and oxides of nitrogen from the polluted gases.

COURSE 4: SOIL AND GROUNDWATER POLLUTION FROM AGRICULTURAL ACTIVITIES

1. Agricultural Activities and Environmental Impacts
2. Soil: Process and Properties
3. Groundwater Hydrology
4. Soil and Groundwater Pollution
5. Irrigation and Drainage: Impact and Remedial Measures
6. Environmental Impacts of Fertilisers
7. Pesticides and Environmental Impacts
8. Monitoring of Soil and Groundwater Pollution
9. Soil and Groundwater Treatment
10. Conservation and Management of Soil and Groundwater

Today, environmental pollution is a major concern all over the world. When sources of pollution are enumerated, agriculture is, with increasing frequency, listed as a major contributor. Agricultural activities can contribute to deterioration in soil and water quality through the release of several pollutants such as pesticides, animal manure, fertilisers, sediments (due to siltation), and other sources of organic and inorganic matter. Hence, in this Course, "Soil and Groundwater Pollution from Agricultural Activities", we will introduce you to major agricultural activities in India and their impact on soil and groundwater. In addition to this, you will be introduced to various remedial measures. In Unit 1, you will learn the basic aspects of agricultural activities. Subsequently, we will introduce you to the soil properties, classification and processes in Unit 2 and in Unit 3 you will learn groundwater characteristics, movement and recharge aspects. This is followed by discussion on soil and groundwater pollution from various sources, impacts of irrigation, drainage, fertiliser and pesticide on soil and groundwater in Units 4 to 7. In the eighth and ninth Units, various sensors/devices used to monitor pollution in soil and groundwater with their remedial measures will be introduced. Finally, conservation and management of groundwater and soil are presented in tenth Unit.

COURSE 5: MUNICIPAL WATER AND WASTEWATER TREATMENT

1. Water as a Resource: Some Related Issues
2. Municipal Waste Water Generation, Conveyance and Disposal
3. Water Treatment Process
4. Wastewater Treatment: Physical and Chemical Processes
5. Wastewater Treatment: Biological Methods
6. Wastewater Reuse
7. Sludge and Septage: Treatment and Disposal
8. Wastewater Disposal into Natural Water Bodies
9. Wastewater Disposal on Land – Land Application
10. Selection, Planning and Implementation Guideline for Water and Wastewater Treatment Systems.

In this Course, "Municipal Water and Wastewater Treatment" we will learn the concepts of water and wastewater treatment. In Unit 1, we will discuss concepts of water availability, quality, attributes, use-based classification, hydrological cycle, etc. In the latter part of the Unit, water quality for various uses, municipal water supply complexities, water leakage have been explained. Unit 2 on the other hand discusses aspects related to municipal wastewater collection, conveyance and treatment. Various components of wastewater collection system have been explained along with concepts such as type of pollutants, mass balance, nitrification, reaeration and dosage. Water supply and treatment processes have been described in Unit 3. Physical and chemical processes involving screening, sedimentation, flocculation, filtration, chemical precipitation, softening and disinfection have also been explained. Unit 4 pertains to physical and chemical treatment processes including ammonia stripping, foam fractionation, reverse osmosis, ion exchange, electrochemical treatment, electro- dialysis etc.

Unit 5 details the biological treatment processes for wastewater. It is important to understand the concepts of biological growth, growth kinetics for aerobic and anaerobic systems as well as practical aspects of biological treatment. One of the important objectives of wastewater treatment is to meet the regulatory requirement; however, in recent times, disposal of treated wastewater is not viewed as an environment friendly option. There is a lot of emphasis on investigating the alternative use of the wastewater for municipal, industrial, agricultural, forest or for recreational purposes. We will discuss these aspects in Unit 6. The treatment processes always produce sludges of different qualities and quantities that need appropriate treatment and disposal, which is discussed in Unit 7. Concepts and standards relating to discharge and receiving water quality related to treated wastewater disposal in lakes, rivers, estuaries and oceans are discussed in Unit 8.

Wastewater application and disposal on land have been discussed in Unit 9. Finally, we will describe in Unit 10 the selection, planning and implementation guidelines for water and wastewater treatment processes.

COURSE SCHEDULE

I. Classroom briefing and introductory sessions at the beginning of the course (2-3 days).
II. Interactive session through Web and Email for three months.
III. Mid-term exam after the third month.
IV. Field Work in the fourth month and final contact sessions for three days.
V. Final exam at the end of the fourth month.

QUALIFICATIONS AND REQUIREMENTS

This Course is open to all graduates. Participants are expected to take part in all activities that form a part of the Course; in addition, the participants should work on individual projects. During the Course, participants should compulsorily attend interactive chat sessions, for which some grading will be given. Participants should actively participate in the sessions for best results. A certificate along with grades will be given to all the successful participants.

EVALUATION

Evaluation will be done at two levels, namely, the concept level and the module level. At the concept level, after student goes through each concept, he/she is evaluated online by randomly displaying multiple-choice questions. Each question carries some weight or marks. Based on his/her test performance, the system checks the number of questions correctly answered and the performance is evaluated. At the module level, after the students complete all the concepts, they are given the final exam, which will be held at the Centre for Ecological Studies (CES)/ Centre for Continuing Education. The concept level evaluation marks will also be considered for grading. Completion of a project is a partial requirement for the Course.

Index